Praise for *Power Shift*

"The great work for present and foreseeable generations is the hard transition from industrial civilization to ecological civilization—and it's all about energy. Given Stayton's compelling and reader-friendly narrative, that would be reason enough to praise *Power Shift*. But there is so much more; namely understanding ourselves, our story, and civilization itself as defined by energy. I cannot recommend this volume strongly enough as our guide to a durable future."

— Larry Rasmussen, Reinhold Niebuhr Professor Emeritus of Social Ethics, Union Theological Seminary

"Robert Stayton does for energy what Michael Pollan did for food. You get the history, the science, the compelling reasons we need to go solar in a book that both interests and inspires. *Power Shift* is the most compelling and clearly written argument for solar power that I have encountered in several years of reading on the subject. Five stars for Robert Stayton!"

—Margaret Payne, Educator; Advisor, Nautilus Book Awards

"The Sun was humanity's primary source of energy in the past and it will be again in the future. *Power Shift* tells the history of people using energy and points the way toward a post-carbon future where we recognize and harness the abundance of perpetual solar power that falls upon us every day. We have been trained as customers by energy companies to be consumers of their product. Distributed solar PV technology will allow us to empower ourselves and our communities. *Power Shift* tells how we can design and build a version of the solar powered future we can all hope to leave to our children and grandchildren."

—Eric Youngren, Founder and CEO of Solar Nexus International

"*Power Shift* provides a thorough review of the history of energy and a compelling vision for the use of solar photovoltaics to power humankind. This would be a good textbook for engineering, physics, and science students in college and university, to complement their other equation-oriented texts."

—Mike Arenson, President of Arenson Solar

"Overall, fantastic! Clear, understandable to the general public, authoritative, free of jargon, and not preachy. I think you hit the nail on the head."

—Dr. Douglas Brown, Physics Instructor Emeritus,
Cabrillo College

"*Power Shift* describes in detail why and how the shift to solar power as our dominant energy source will happen in the US and the world. But it also describes how the relationship between human civilization and energy has evolved, from our original use of fire to our current time-limited reliance on fossil fuels. Stayton avoids technical jargon but ably shows how to assess and compare quantitatively, our past, current and possible future energy sources. This way, he moves readers well beyond the partisan bickering and dithering that has effectively stalled the national discourse on our energy future. Bravo! Please read this well written, informative and timely book."

—Doug McKenzie, solar analyst and founder of
Lights On Solar

"I think this book is a landmark document at a critical historical juncture. It's a terrific primer for anybody seeking solid basic grounding in energy literacy. It makes clear why any citizen at this moment in history should cultivate such knowledge, and provides it in a painless and enjoyable way."

—Sarah Rabkin, Teacher~Editor~Writing Coach

"We all know we have to stop burning fossil fuels. But how to kick the habit? Robert Stayton has written an inspiring book about how solar can take us smoothly into a fossil free future. *Power Shift* clearly explains how solar energy can replace fossil energy, interweaving all the science the average reader needs to follow along. Stayton's approach is thorough, user friendly, and firmly grounded in current science."

—Jennie Duscheck, Freelance Science Writer

Power Shift

From Fossil Energy to Dynamic Solar Power

Robert Arthur Stayton

Illustrated by Todd Sallo

Sandstone
PUBLISHING

Sandstone Publishing
PO Box 2911
Santa Cruz, CA 95063
info@sandstonepublishing.com
www.sandstonepublishing.com

ISBN 978-0-9904792-0-8 (Print edition)
ISBN 978-0-9904792-1-5 (Kindle edition)
ISBN 978-0-9904792-2-2 (EPUB edition)

Library of Congress Control Number: 2014922518
BISAC Subject Code: TEC031010 TECHNOLOGY & ENGINEERING /
    Power Resources / Alternative & Renewable

Credits

Book design by Sandy Bell Design and Robert Arthur Stayton.

Cover design by Robert Arthur Stayton and Sandy Bell Design.

Cover photo of the sun taken in 304 Ångström light (extreme ultraviolet) imaged by the Solar Dynamics Observatory's Atmospheric Imaging Assembly (AIA) instrument, courtesy of the US National Aeronautics and Space Administration, downloaded 12 August 2014 from http://svs.gsfc.nasa.gov/vis/a010000/a011200/a011298/20130621_025956_4096_0304.jpg

Human-powered medieval winch image (page 31): Marie Reed - Own work. Licensed under Public domain via Wikimedia Commons - http://commons.wikimedia.org/wiki/File:Treadmillcrane.jpg#mediaviewer/File:Treadmillcrane.jpg

Millpond illustration (page 35) by Todd Sallo, inspired by *Mills on the Tsatsawassa: Techniques for Documenting Early 19th Century Water-Power Industry in Rural New York*, by Philip L. Lord, Purple Mountain Press, Fleischmanns, New York, 1983.

Newcomen engine illustration (page 41) by Todd Sallo, inspired by: *Science for the Citizen: A Self-educator Based on the Social Background of Scientific Discovery* by Lancelot Thomas Hogben, W.W. Norton, 1956.

Factory powered by a single steam engine photo (page 43): Gebr. Steimel GmbH & Co., 53773 Hennef, Deutschland, unaltered, originally titled Foto der Dreherei der Gebr. Steimel GmbH & Co., 1905. Licensed under the Creative Commons Attribution-Share Alike 3.0 Germany license, via Wikimedia Commons - http://commons.wikimedia.org/wiki/File:Steimel_Produktion_Dreherei_1905.jpg

Abengoa Solar PS20 concentrating solar power plant photo (page 145): PS-20 concentrating solar power plant. Sanlúcar La Mayor (Seville) Spain (technology owned by Abengoa Solar, S.A.) ©Abengoa Solar, S.A. 2015. All rights reserved.

Declining oil discoveries illustration (page 86) by Todd Sallo, created from data provided by ASPO International, www.peakoil.net.

All other illustrations by Todd Sallo.

This book uses material from several Wikipedia articles as noted in the end notes. These articles are released under the Creative Commons Attribution-Share-Alike License 3.0, http://creativecommons.org/licenses/by-sa/3.0/.

Dedicated to
My dear wife Mary Tsalis
and
My father Chester Arthur Stayton, Jr.

Contents

Preface

It seems impossible that we are not facing up to the greatest threat in modern times to the survival of human civilization. The disruption of climate patterns caused by carbon dioxide from burning fossil fuels menaces all people and all nations. Our warmed seas are rising and already flooding low-lying island nations, with all the coastal cities of the world waiting next in line. Glaciers and snow packs from California to India are melting away, inevitably drying up the water sources that millions of people depend on. Rainfall patterns that sustained farming for centuries can no longer be relied upon, threatening food insecurity and political instability.

Now the threat has doubled, because the extra carbon dioxide sinks into our oceans and makes the water more acidic. Shelled animals can't form their shells in acid water, killing them off before they can grow, and killing off all the animals that depend on them for food. We teeter on the edge of mass extinctions from which we cannot recover.

Yet our carbon emissions continue to grow, because world leaders are doing little to control the burning of fossil fuels that are causing these problems. After decades of negotiations, no global treaty exists to control carbon dioxide. Developing nations accuse the already-developed world of creating the problem, but the developed world won't act unless the developing nations sign on to the same restrictions. We seem to be waiting for a sign, a sudden event so catastrophic that it compels us to overcome these differences and take action. But this disaster does not rush upon us like a hurricane, but instead slowly erodes our foundation like termites. If we wait until the house collapses, we will have waited too long.

We need not wait. I wrote *Power Shift* to shout from the rooftop that WE ALREADY HAVE A SOLUTION TO GLOBAL WARMING. That solution comes in the form of solar panels and wind turbines that derive their energy from the sun. When combined with energy storage, they form a clean energy system that can completely replace fossil fuels. We can gradually phase them in so that in fifty years we will be 100% solar powered. And because solar energy does not run out like fossil fuels, we will never need to go looking for another energy source again.

Power Shift provides a complete blueprint for reaching this future. More importantly, it describes how you can join in, and why you would want to. These twin global crises derive from how we use energy, and everyone uses energy. But today's energy debate is jammed in a deadlock between energy and the economy. This narrow debate has failed to notice that the world is undergoing a fundamental shift in *how* we use energy.

In fact, the world is transitioning from energy to power, a change that is setting humanity on a new course for the future. This shift is so fundamental that it defines a new historical epoch, and looms as large as the discovery of fire or the exploitation of fossil fuels. Today few are aware that this new epoch has already begun. *Power Shift* is the first book to describe this underlying change in the human relationship with energy.

Power Shift explains this insight by retelling the story of humans in terms of energy. It shows in plain language how energy defined our species and supported our takeover of Planet Earth. It clearly outlines our current energy dilemma, and the way forward.

If you live a modern energy-enabled lifestyle, then you are a participant in this crisis, and you have choices to make. The energy decisions we make today will decide the fate of many generations to come. This book raises the level of energy debate to match this historic significance.

Power Shift will enlarge your view of the world through the lens of energy. From that new vantage point, you will see hope for the future.

How We Got Here

CHAPTER 1

Energy Defines Us

You have probably forgotten you are an animal. More specifically, you are a mammal of the species *Homo sapiens.* I can state that fact with confidence because only *Homo sapiens* read books.

You were born of a mammalian reproductive system, and you will die when your animal body gives out. Like other animals, you must breathe air, eat food, and take sleep to stay alive but you probably don't think of yourself as an animal because your life carries on so differently from the lives of other animals.

If you lived the life of an animal in the wild, you would spend most of your day searching for food, while keeping up a constant vigilance against predation, which is, of course, someone else's search for food. Do you spend your day searching for food? Probably not. Do you fear becoming food for someone else? Not likely. Why are our lives so different from those of other animals?

In a word: *energy.*

That's probably not the word you were expecting. Most would consider our complex brain to be the most significant feature that distinguishes us from other animals, followed closely by our hands with grasping thumbs, and our language skills. Our advanced brain allows us to comprehend our environment and imagine how to change it. Our brain guides our hands in making tools that extend our capabilities, and enables our language skills that let us pass on those capabilities.

These assets are necessary but not sufficient to explain how we ascended to our current modern life. The crucial difference emerged when we used these skills to control energy. When we learned to control fire, our first energy source from outside of our bodies, we permanently parted ways with the animal lifestyle.

Our continued development has tracked a continuous escalation in our control of energy. It was energy that built modern society, and it is energy that runs it. Without energy, we would be just a talkative primate that walks upright on the savannah, hunting and foraging like other animals.

With energy, we have built homes that provide security and comfort, food-production systems that feed us reliably, and transportation systems that can carry us anywhere in the world. If all of our energy devices were taken away, all of this would grind to a halt. All manufacturing would stop, so there would be no goods to buy. All transportation would halt, so we would have to walk or ride horses. And all food systems, from farming to supermarkets, would cease operating, so we would have to relearn primitive survival techniques. Imagine what you would do if no food were available in any store. The food is there because of energy.

Energy is the key factor that separates us from other animals. Our ability to direct the flow of energy is a unique, defining characteristic of humans. We are *the energy-using animal.* As the energy-using animal, we are in a class by ourselves. We don't observe squirrels building fires to keep warm, or chimpanzees driving around in cars. Only humans have mastered the skills to manipulate energy to serve our purposes.

"Energy is the key factor that separates us from other animals."

To be clear, all animals use energy in the form of food. Food contains chemical energy that powers the bodily functions that keep all animals alive. The human body has a digestive tract similar to those of other mammals, through which we derive the energy to power our hearts, minds, and muscles.

But for other animals, food is their *only* energy source, and food-powered muscles are their only means of accomplishing anything.

When a wildebeest on the savannah needs to migrate to find fresh grass, it has no option but to walk there, powered by its food energy.

Humans have learned to direct *external* energy sources—energy from outside of our bodies—in quantities that far exceed the energy available in food. If we needed fresh grass, we would use gasoline energy to drive there, or better yet, have it cut, packaged, and express delivered. Using external energy multiplies our capabilities beyond muscle energy, enabling us to manipulate our environment like no other animal.

The advantages humans accrue from energy have evolved into an ongoing relationship. Our relationship with energy has many facets, but it can be summarized as follows: we direct the flow of energy, and in turn, energy grants us great power over our environment.

Few would doubt that humans have altered our planet, but how many realize that energy is the dominant factor behind those changes? Because energy is the agent underlying every action, every time we use energy we change something around us. The more energy we use the bigger the changes we make. Because we have the ability to control energy, we direct those changes in ways that for the most part benefit ourselves. That control defines our relationship with energy.

This relationship with energy started before we were fully human, when our prehuman ancestors learned to control fire to alter their local environment to survive. As our energy-using skills developed over time, we were able to safely direct ever-larger quantities of external energy to accomplish ever-greater tasks. Modern energy-powered machinery routinely levels mountaintops and erects skyscrapers. With energy, we have even shot the moon, using liquid hydrogen energy to lift twelve Americans to the surface of the moon.

With energy supporting us, our species has multiplied and spread over the surface of the earth. In cold latitudes, we use energy to keep us warm, and in warm latitudes, we use energy to keep us cool. As we spread, we take over land, clearing it for farms, roads, and cities, and chasing off all the wild animals. The few wild areas that have not yet been taken over by humans continue to shrink as our population grows.

Our wonderful mental abilities have enabled us to gradually recognize the effects of our planetary takeover. As we have studied and catalogued plant and animal species, we have noticed that the numbers of many wild species are dropping, sometimes to extinction. So we put in place species-preservation programs, investing time and energy to preserve wild habitat, and even resorting to controlled breeding when the numbers get too low. When the numbers get too high, we thin the herd, acting in place of the predators we most likely eliminated.

We have shifted from being just residents of Earth to being managers of Earth. We effectively control life on Earth, determining which species live or die. We like to think there are still wild places, but most are preserves that humans have set aside, circumscribed by boundaries that we establish, containing populations that we monitor. If other animals could develop energy-using habits, then they could compete with us, but they will not have that chance while humans dominate.

All this control derives from energy—massive quantities of energy. These massive quantities are actually a relatively recent development. Until the early 1800s, wood fire and animal muscle provided most of our energy, and both were self-limiting: wood can be harvested only as fast as a forest can regrow, and animals take up land and food.

The discovery of large caches of stored energy in the ground released us from those limits. Coal, oil, and natural gas are the fossilized remains of plants and animals that lived millions of years ago. When they died, their decay process was somehow interrupted, so part of the energy stored in their body tissues was preserved and condensed, gradually becoming the fossil fuels we use today.

With the development of coal, oil, and natural gas over the last 200 years, our energy use became limited only by how fast we could pull them out of the ground and find new ways to burn them. We replaced dwindling wood supplies with seemingly unlimited coal in lime kilns to make cement and in foundries to make steel. We learned to burn fossil fuels in boilers to make steam to generate electricity, and we learned to burn fossil fuels in engines to power cars, trucks, boats, trains, and airplanes.

Because each new use added more demand, the fossil-fuel industries grew almost continuously for 200 years. It is hard to imagine that today we burn 90 million barrels of oil *per day* worldwide, and that oil is just one of *three* fossil fuels that we burn in massive quantities. The word "massive" hardly seems adequate to describe our energy use.

Our marvelous brains also began to note the effects of all this burning. Initially we noticed the ugly air pollution that clouded our cities. We responded by applying technology to clean up the combustion process to emit fewer pollutants. Cars were fitted with catalytic converters, and coal plants were fitted with scrubbers. The skies over many large cities cleared.

Now we face another side effect that is proving much more difficult to correct: Carbon dioxide. When fossil fuels burn, the carbon in the fuel combines with oxygen in the air to form carbon dioxide, an odorless, colorless, nontoxic gas. The carbon in fossil fuels comes from those ancient plant and animal remains, and it carries the energy that we seek. Since forming carbon dioxide is necessary to getting at that energy, there is no simple technology to turn off the carbon dioxide produced by burning a fossil fuel.

The release of carbon dioxide had previously been considered harmless because it is chemically inert in the atmosphere, unlike other pollutants that undergo smog-producing chemical reactions. Carbon dioxide has always been a natural component of the atmosphere that you inhale with each breath. Your own body produces it as a natural waste product, which you exhale with each breath.

The problem comes from the *scale* of carbon dioxide emissions from fossil fuels. Worldwide we pump about 100 million tons of carbon dioxide into the atmosphere *every day*. In equivalent volume, we burp a bubble of carbon dioxide three miles in diameter every day.[1]

And unlike other air pollutants, carbon dioxide stays up there because carbon dioxide is chemically inert in the atmosphere. The natural cleansing mechanisms that remove extra carbon dioxide from the atmosphere work much more slowly than the rate of our additions, so the carbon dioxide accumulates over the years.

This accumulation is proving to be a problem because of a simple physical property of carbon dioxide—it absorbs infrared radiation. Infrared is the same as light but beyond the visible red portion of the spectrum. An infrared camera shows outlines of warm bodies glowing with infrared light energy. Warmed by the sun, the earth's surface glows in the infrared range. The extra carbon dioxide in the atmosphere blocks some of the outgoing infrared energy that would normally escape into space.

The extra carbon dioxide effectively turns the Earth into a giant greenhouse, with the atmosphere acting as the greenhouse glass that

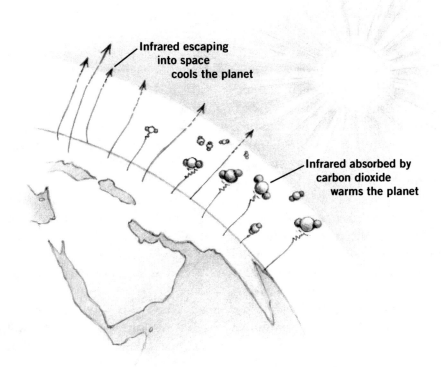

Figure 1. Carbon Dioxide absorbs outgoing infrared

The infrared radiation that escapes the atmosphere carries away energy, cooling the planet. But some of the outgoing infrared energy is intercepted and absorbed by carbon dioxide, thereby warming the atmosphere.

retains heat and raises the Earth's temperature. The resulting global warming melts ice, raises ocean levels, and alters climate patterns. The evidence is all around us.

But it gets worse. About a fourth of all the carbon dioxide emitted from fossil fuels over the last 200 years has been absorbed into the world's oceans. That seemed to be good news at first, because it slowed the accumulation of carbon dioxide in the atmosphere, but it is proving to have disastrous consequences for marine life.

The carbon dioxide combines with seawater to form carbonic acid, which makes the ocean water more acidic, according to the Ocean Acidification Fact Sheet from the National Oceanic and Atmospheric Administration (NOAA).[2] The oceans have no ready mechanism to neutralize the extra acid, so it accumulates. Shelled creatures have trouble forming shells in acidified water, because their shells dissolve as fast as they can make them. Many of the tiny shelled creatures form the base of marine food pyramids, so the acid threat to them threatens species at all levels. Ocean acidification is now seen as "an equally evil twin" to global warming, says Jane Lubchenco, head of NOAA.[3]

Our fossil-fuel-based energy systems have grown continuously for so long that they have now grown very large, so large that their carbon dioxide emissions have modified the planet in two major ways. In the period from 1800 to 2014, our fossil fuel burning has increased the carbon dioxide concentration of the entire volume of the Earth's atmosphere, all 12 billion cubic miles, by 43%.[4] During that same period, the carbon dioxide that was absorbed into the oceans has increased the acidity of all the world's oceans by 30%.[5]

Read those percentages again and let their significance sink in. These are not local changes affecting a few people; they are global alterations of our air and water, astonishingly large changes perpetrated by a single species, the energy-using animal. Since air and water envelop all life on Earth, such changes are altering the patterns of life on Earth, including the lives of humans.

We have now come to realize that the energy systems that established human control over life have unintended consequences that are spinning out of control. In a sense, we have developed an unhealthy relationship with energy: the more we use fossil fuels, the more harm we do.

Yet despite worldwide acknowledgment of the problems, little is being done to arrest these changes. The first and only global treaty to restrict carbon dioxide emissions was the Kyoto Protocol adopted in 1997. It expired in 2012 without a replacement. The treaty's purpose was to reduce the rate of carbon dioxide emissions, but during its lifetime, that rate accelerated by 41% worldwide.[6] The ineffective Kyoto Protocol may be extended, but there is nothing to replace it. Attempts in Durban, South Africa, in 2011 and Doha, Qatar, in 2012 resulted only in promises to develop a new treaty by 2015, and if that is successful, to begin enforcing it in 2020.

If you feel powerless in the face of such international foot dragging, you are not alone. Fortunately, the top-down approach is not the only way to correct our course. A bottom-up approach is not only possible, it is far more likely to succeed, as you will see in later chapters. That's because carbon dioxide emissions are primarily an energy problem, and *everyone* uses energy.

If you live a modern lifestyle, then you are a player in this drama. When you drive a car, its exhaust contributes carbon dioxide to the atmosphere. When you turn on a light or use an appliance, the power plant that generates your electricity will burn a bit more coal or natural gas and emit a bit more carbon dioxide in your name.

Simply asking people not to drive and not to use electricity to avoid carbon emissions will not work, because such energy use is necessary, not optional. Wasteful energy habits can be corrected, but there is still a baseline requirement for significant amounts of energy to live as modern human beings.

Yet our energy requirements do not have to tear up the planet. We can gradually substitute energy sources that are free of carbon emissions. We have such energy sources today, most of which derive their energy from the sun. Solar panels, wind turbines, and other renewable energy sources, when combined with energy storage, can form a complete energy system for our future—one that emits no carbon dioxide.

It remains an open question whether we will deploy these new improved energy systems in sufficient quantities to make a difference. Those who know the most about energy say we must, while those who control political and economic power say we cannot. If this impasse is not broken, we will drift into a future not of our choosing but of our inertia. If we want to avoid unintended consequences, we need to establish a set of *intended* consequences, and work toward them as a civilization.

The good news is that despite the inability of world governments to agree on a path forward, the transition to clean solar energy is already proceeding at a rapid pace. The greatest progress is being made by individuals, businesses, and organizations deciding for themselves to switch to carbon-free energy. You have that choice too. It is actually easier for you than for big governments and energy corporations.

Why? Because solar energy is all around us. You have direct access to an exceptional energy source. It is remarkable that we have failed to see the solution that is falling on our heads every day.

That failure comes primarily from our limited view of energy as a consumer item, like soap or potatoes. We buy energy, consume it, and then buy more. The concept of energy as a consumer item is relatively recent in our long history as the energy-using animal. It grew concurrently with the growth of fossil fuels, a form of energy that can be metered and sold. If we are going to get past fossil fuels, then we will have to get past the simplistic view of energy as just a consumer item.

This book will expand your view of energy and your view of the world through the lens of energy. We start by re-envisioning our past in terms of energy to answer the question: How did we become the energy-using animal?

PART I: HOW WE GOT HERE

CHAPTER 2

How We Became the Energy-using Animal

Humans did not become the energy-using animal through natural selection. Our energy-using habits are learned behavior, not a genetic mutation. Through trial and error, our ancient ancestors discovered ways to exploit energy to make their lives better. Those discoveries were developed by advanced brains and skillful hands, and passed along to others through example and language.

Since you were not born with genes to operate a light switch or drive a car, you had to be taught those skills. As you were growing up, you learned from your parents, teachers, and others how to operate energy-using devices. From this training, you developed your own personal relationship with energy. That relationship makes your life more secure, more comfortable, and more interesting.

Each generation passes along its energy-using knowledge, and some generations add to it by developing new energy systems. That chain of learning extends back unbroken through thousands of generations, connecting you with our earliest ancestors teaching their children how to keep a fire going.

The story of how we became the energy-using animal is epic, beginning before we were fully human, spanning our long development into Homo sapiens, and dominating all 5,000 years of our recorded history.

Although energy played a fundamental role in shaping our development, few recognize that role. Energy is so pervasive in our lives that we rarely notice it. Everything you do every day involves energy of some kind. Walking, cooking, driving, all require energy. Even sitting quietly reading a book requires energy to keep your heart pumping and your brain synapses firing.

Just as a fish is unaware of the water it swims in, so are we humans generally unaware of the energy that powers our lives. Energy is so ever-present that you notice it only when it is lacking, as when the power goes out.

To help you appreciate the depth of your relationship with energy, it helps to bring energy into closer focus. To that end, I introduce the *energy scope*, a device of my own invention for focusing on energy.

Every reader of this book is entitled to their own energy scope. You can use it to highlight energy, to frame it in your field of vision so you can see what it does. As you might have guessed, my energy scope is a literary device, not a physical one. Think of it as a tool for aligning your imagination with the reality of energy.

Using an energy scope means viewing and interpreting an event in terms of its energy flows. With your energy scope, you can focus on your everyday life to see how you depend on energy for your safety and comfort and your very survival. An energy scope can zoom in like a microscope to examine photosynthetic energy flows in plant cells, or zoom out to survey the vast landscape of global energy systems.

An energy scope gives you a new view of the world you live in, a view from an energy perspective. It will widen your understanding so you can see the essential role energy plays at all levels, far beyond its role as a commodity for purchase and consumption. That broader understanding will be useful when choosing future energy sources.

When We Were Animals

In the time before we learned to use energy, we lived as other animals did. Imagine yourself as one of our primitive ape-like ancestors, living in a time *before* stone tools and fire. Without fire to protect us at night, we slept in trees. Without stone tools, we foraged rather than hunted for most of our food.

Living the life of animals in the wild, we survived by paying attention. We had to remain wary at all times, using all of our senses to alert us to danger. Two of those senses, hearing and sight, could detect action at a distance by tuning in to natural energy flowing in our environment.[1]

When viewed through an energy scope, the act of hearing can be seen as a transfer of energy from some source to a receptive ear.

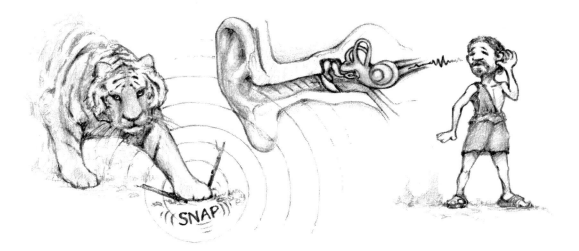

Figure 2. Sound transfers energy to the ear
When a walking animal steps on and breaks a twig, some of the vibrational energy of the snapping twig transfers to the air, vibrating the air. Because air is elastic, its springiness transmits the energy rather than absorbing it. So the vibrational energy spreads outward from the source in all directions, traveling on the elasticity of the air. If you are nearby, some of that traveling energy enters your ear. The eardrum acts as an energy detector, converting the tiny air vibrations into electrical nerve signals that are sent to your brain, which tries to interpret the sound as prey or predator.

Likewise, the act of seeing can be interpreted as a transfer of light energy from a source to a receptive eye. Light is a different form of energy from sound, so the receptors in your retinas differ from the receptors in your ears. The effect, however, is similar. The optic nerves transmit signals to your brain, where your mind builds a mental picture of your surroundings.

Without knowing how they do it, all animals exploit these natural energies of sound and light to survive. Those animals that could hear and see better than others survived the longest and produced the most offspring, so natural selection over many generations evolved animals with exquisite sensitivity to tiny amounts of energy.

Our species inherited and refined these mechanisms. With our forward-facing eyes, we could detect depth and distance—important information for surviving in the wild. With the addition of full-color vision, we could distinguish predatory animals from green vegetation. Hearing was probably even more important for survival, because you can hear sounds from behind you, outside your field of vision. If our hearing were any more sensitive, we would pick up noise from the random fluctuations in air pressure due to air molecules banging together.

These sensitive nerve mechanisms are powered by chemical and electrical energy in the body. In fact, all your bodily processes require energy to keep operating, and that energy derives from the food you eat.

When viewed through an energy scope, food can be seen as a source of stored chemical energy. Because food comes from complex living things, the energy is embedded in a jumble of organic compounds. An animal's digestive tract must break down and sort out these compounds into useful components, not unlike how an oil refinery processes crude oil. Your gut converts the food into chemical forms that can circulate in the blood to all cells in the body. That continuous flow of chemical energy keeps the cells alive and functioning. If the energy-carrying blood is cut off to part of a body, that part will die.

The most visible manifestation of the energy derived from food is movement, made possible by our muscles. Muscles convert chemical energy directly to motion, a trick not yet duplicated by modern science.[2] Without muscles, we would be as immobile as plants. With muscles, we walk, run, leap, and climb. In our early days, we used our muscles to gather food and flee predators.

The standard unit of food measurement is the Calorie, which is actually a unit of energy measurement. A Calorie counter is like the meter on a gas pump, measuring the flow of energy into your body. Your amazing body runs on the food equivalent of a cup of gasoline per day.[3]

When you eat, you are partaking of energy, thus refueling your body. This act is so important that you usually plan your day around your fuel stops. If you don't always know *where* you will stop to eat, you generally know *when*, typically three or more times per day. If you forget, your body conveniently reminds you.

"Your amazing body runs on the food equivalent of a cup of gasoline per day"

Where does the energy in our food come from? Without exception, that energy comes from the sun. Plants convert light energy from the sun into chemical energy, which the plant stores and uses as needed. When a cow eats grass, it ingests chemical energy stored in the grass's tissue and stores some of that energy in its own body. When you eat a hamburger, you get some of that energy for your own use. In a real sense, you are eating solar energy, twice removed.

Plants have evolved the special ability to convert the energy in sunlight directly to chemical energy through photosynthesis. Animals can't do that; the only way an animal can directly use the sun's energy is to act like a lizard and bask in sunlight to get warm.

When you zoom in on a plant's leaf with your energy scope, you find that chlorophyll pigments in the plant's cells use the sun's energy to drive a complex sequence of chemical reactions. These combine carbon dioxide from the air and water from the ground into glucose, a simple sugar. The chemical bonds in glucose contain more energy than those in carbon dioxide and water, effectively storing the absorbed solar energy in a stable chemical form.

A plant can release some of its stored energy as needed for its own metabolism. If an animal eats and digests the plant, it can extract some of that energy. When the energy is applied, the glucose reverts back to carbon dioxide and water. You exhale the carbon dioxide and excrete the water to complete the cycle.

Since all food is either plant matter or animal matter, that means all your food energy derives from solar energy.[4] The food energy that keeps your heart beating, your nerves reacting, and your muscles flexing all comes from the sun. That makes you a solar-powered being. With few exceptions, all life on Earth is 100% solar powered.[5]

Hunting Weapons

If you go back to the time before we had weapons, our ancestors had the same relationship with energy as other animals. We used sound and light energy to sense our environment, and we used food-powered muscles to move about to obtain enough food for survival. We were mostly vegetarians because obtaining meat would have been difficult. Humans are not particularly fast runners or great leapers compared to other animals. We have no sharp claws, and our teeth could not be called ferocious. Imagine trying to kill or capture live prey with your bare hands.

Our first breakthrough toward becoming the energy-using animal was the discovery of kinetic energy, the energy of matter in motion. You can imagine a smaller hairier version of ourselves picking up a stone to throw or a long bone to swing as a club.

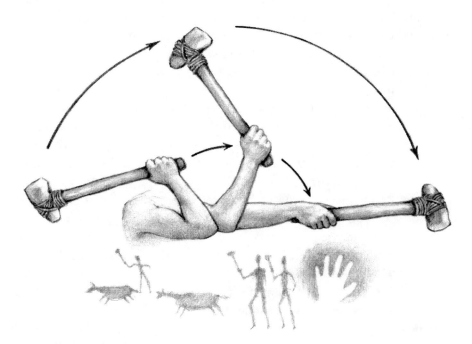

Figure 3. Kinetic-energy weapon

When you swing a club, the end of the club travels a greater arc than your hand and achieves a greater speed. So the tip of the club carries more kinetic energy than your hand because it is moving faster. If the end of the club is a stone, then it weighs more than your hand, and that also gives it more kinetic energy.

The word *kinetic* derives from the Greek word for motion, also the root of the word *cinema* to describe moving pictures. Whenever matter is put in motion it has kinetic energy. It takes both matter and motion to make kinetic energy.[6]

When viewed through an energy scope, a club can be seen as a kinetic-energy device. The kinetic energy comes from your muscles, of course, but the device focuses that kinetic energy to better effect.

While the first clubs were probably large animal bones or sticks found on the ground, around 2 million years ago our early ancestors learned to work stone. At some point they attached stones to sticks, and put a sharp edge on the stone.

When the stone end of the club collides with living flesh, the stone (being an inelastic material) does not absorb the kinetic energy but instead transfers the kinetic energy to the flesh, and concentrates it on a small area of contact. If the stone has a sharpened edge, then the area of contact is very small. Applying a lot of energy to a small area will cause damage to living tissue, which is the whole point of the hunting weapon. If the damage is extensive enough, it disrupts the prey animal's living processes and disables or kills it.

Without naming it, we had discovered kinetic energy. Modern chimpanzees have been observed swinging clubs, but they don't use them on a regular basis to obtain food. The more advanced brains of our ancestors recognized the value of directing kinetic energy, and passed that practice on to others.

The discovery of the club energy system made getting food easier and safer. By directing their muscle kinetic energy more effectively, early humans improved their chances of survival.

Weapons also probably altered their physical development. By enriching their diet with meat, they had sufficient fuel to support growing the human brain to its present size (your brain consumes 20 percent of your food energy despite being only 2 percent of your body weight).

The Fly Hunter

Swatting a fly is how modern people carry on the ancient tradition of swinging a club. A flyswatter is certainly a featherweight as a club, but it is still a kinetic-energy weapon that is deadly to flies. With flyswatter in hand, you stalk your prey to keep your life clean of flies, similar to your ancient ancestors' stalking of larger prey in pursuit of food.

A spear is a step up from a club as a kinetic-energy weapon. A spear has a sharpened point. That means the kinetic energy is concentrated on an even smaller area, the contact area of the spear point. That high concentration of kinetic energy is sufficient to tear skin, permitting the energy to be delivered deeper into the body where it would be even more disruptive to life-support systems.

A spear also has a longer shaft, which allows the spear carrier to keep a bit of distance from potentially dangerous prey. Even better, a spear can be thrown from a distance, providing even greater safety (assuming you don't miss).

The bow and arrow introduced a more complex kinetic-energy system. Pulling back on a bow stores elastic energy in the bent bow. When released, that elastic energy is transformed into kinetic energy of the arrow. By releasing stored energy, the arrow can achieve a greater speed than if thrown by hand. Greater speed means more kinetic energy, so it can travel farther and penetrate deeper.

The idea of a bow and arrow may have derived from the use of snares to catch small animals.[7] When setting a snare, a hunter uses some of their muscle energy to bend a branch to store elastic energy. When an animal triggers the snare, the elastic energy is released as kinetic energy to close the snare. By employing stored energy, hunters could set many such snares at once, increasing their odds of capturing game.

Kinetic-energy hunting weapons made up for our lack of speed and sharp claws in obtaining food. With these weapons and coordinated attacks, no animal was safe from human predation. Lions and tigers and bears were no match for well-armed humans. It is likely that our early ancestors helped wipe out woolly mammoths and other large animals. Thus began the takeover of Planet Earth by the human species.

CHAPTER 3

The Wood Energy Epoch

While stone weapons were our first application of energy, that energy still came from our own bodies. We used muscle energy powered by food, just as other animals use muscle energy for hunting. The real breakthrough in the human relationship with energy came when we engaged energy sources from *outside* our bodies to meet our needs.

Our earliest instance of using an external energy source was probably fire. While stone tools date from 2 million years ago, the first evidence for human-controlled fire is from about 600,000 to 800,000 years ago, when we were still archaic humans[1] (modern Homo sapiens date from about 200,000 years ago).

Fire marked the beginning of the first major epoch in the human relationship with energy. In the study of history, a segment of time with its own distinguishing characteristics can be called an *epoch*. The Wood Energy Epoch began with our first use of fire, and lasted until fossil fuels supplanted wood energy in the 1800s. Along the way, wood was supplemented by energy from animals, wind, and falling water.

Fire

Natural fires started by lightning must have been known to early humans, and, like other animals, humans must have feared them. Death and destruction are easily observed in the aftermath of a forest fire, and smoke is repulsive, even toxic. At some point unrecorded in time our ancestors overcame that fear and learned to control a remnant of a natural fire. Control meant containing the fire so it did not spread, and feeding fuel into the fire to keep it burning. Before we learned to start fires from scratch, it was vital to keep the adopted fire continuously burning.

Unlike a stone tool, fire is not a permanent thing. Fire is transient, requiring periodic feeding or it will go out. When viewed through an energy scope, a simple fire is revealed as an energy *process*, a transformation of wood chemical energy into heat and light energy. The heat and light flow outward from the fire and are eventually absorbed into the environment. That constitutes an active flow of energy, not a static state. Unless the energy outflow is matched by wood-energy input, the fire will exhaust its fuel and die out.

With fire, our relationship with energy began to grow in directions that other animals could not match. Since no other animal learned to use fire or other external energy sources, we became the one and only energy-using animal. That distinction continues to today.

Fire meant life after dark. Without fire, animals huddle down after dark, staying still, and hoping not to be discovered by a night-stalking predator. With fire, early humans experienced a safe new life at night. The flames of a fire provide light energy, so humans could work on tools or clothing in the evenings. The light also provided safety, because animal predators avoid fire. With fire, we went from spending nights in trees, shivering in the cold and dark, to sleeping on the ground, warm and safe in the firelight.

In addition to heat and light, ancient humans also turned fire into a cooking energy system. In cooking, we direct heat energy into food to break it down chemically. That makes the food more digestible, thus making more food energy available. Cooking required hands, eyes, and a brain to control the flow of energy, because too much heat will burn up the food, making it useless for eating.

Figure 4. Fire is an energy process
The keeper of a fire manages an energy flow by collecting firewood and adding it gradually into the fire. Those efforts are rewarded with controlled heat and light to make life better and more comfortable.

Over time, fire moved indoors into fireplaces and wood stoves, which made fire safe to use inside flammable structures. Making light with fire also became more specialized. Around 40,000 years ago someone figured out that animal fat and a wick could provide a long-lasting steady light. A candle is still a fire-energy system, but a tiny fire used only for lighting, not heating.

Brighter lighting came later with lamps fueled with camphene or whale oil. Camphene was a mixture of turpentine from trees and alcohol fermented from grains, and was explosive if not handled with care.

By comparison, whale oil was seen as a premium lighting fuel because it could produce a bright, safe light. Whale oil was probably the first global-scale energy industry. A whale-oil ship would travel great distances to harvest whales for their oil. In 1820, the Massachusetts-based whale ship Maro hunted whales off the coast of Japan, signifying the beginnings of the global reach of energy commerce.

Around 6000 BC, fire enabled another major step forward for human development—metallurgy. When certain rocks were heated in a hot fire, copper, tin, and later, iron could be extracted as liquid metal to be formed into sharp tools and weapons. Metal tools became possible when humans improved their fire-energy system in two ways: by confining the fire to concentrate the heat, and by forcing air into the fire to achieve higher temperatures.

Early metalworkers also discovered that replacing wood with charcoal produced stronger metal by reducing impurities. When wood burns, the heat drives out volatile chemicals that flame. When the volatiles stop coming, what's left is mostly solid carbon—charcoal. Charcoal has the same energy content as coal, is virtually smokeless and sulfur-free, and can be used indoors (vented for carbon monoxide). Making charcoal was inefficient, though, wasting 60 percent of the wood's chemical energy. Starting in the late Middle Ages, demand for metals led to extensive deforestation in Europe and Asia because about 480 pounds of wood had to be burned to make each pound of iron.[2]

Metal tools proved to be far superior to stone tools in almost every way. Metals were less brittle than stone and could be repeatedly honed to a razor-sharp edge. The ability to form arbitrary shapes with liquid metals in molds opened up endless possibilities for new tools.

Metals were essential for all the energy systems that humans developed since that time, either as a material part of an energy apparatus or in tools to make that apparatus. So our first energy system, fire, enabled the development of all of our other energy systems down through the ages.

Agriculture

Even before metals were developed, agriculture energy systems began to appear in various locations. Agriculture qualifies as a human energy system because we have a hand in directing the flow of energy that converts sunlight into food energy. Of course, the basic photosynthetic energy process that plants use to transform solar light energy into chemical energy predates humans.

Humans made food plants their own energy system when they applied muscle energy to improve the flow of solar energy to their food plants. Those efforts included providing sufficient water for sustained growth, removing shade and weeds, and protecting against crop-eating insects and animals. Agriculture can be seen as an optimization of a natural energy system by humans to make more food energy available.

Agricultural fields provide far more food than do areas of wild growth, where only a fraction of the vegetation would be suitable for human consumption. Agriculture permitted more people to be supported by the land. By removing the need to wander to find wild food, people could settle down in one place. This led to permanent settlements, which eventually led to cities and civilization.

The high productivity of agriculture meant that not everyone had to be a farmer. Some of those freed from farming became artists and merchants, while others became tinkerers and observational scientists who went on to develop other energy systems. Agriculture enabled all of these essential starting points for human civilization:[3]

- Stationary settlements;
- Time for making a variety of goods;
- Accumulation of goods;
- Development and accumulation of knowledge; and
- Social hierarchies.

Animal Power

Humans also directed the muscle energy of domesticated animals to ease the burden of field work, turning them into human-controlled energy systems. Animals were used to turn the soil, grind grain, and pump water, all tasks that wear people out if they do them manually.

A large animal can do the work of five to ten people. By delivering higher power, animals enabled the farm work to proceed more quickly and with fewer people. The human role switched from pulling the plow to guiding the plow animal. The "energy-using animal" became the "animal-using animal."

Of course, farm animals require their own energy inputs, so a portion of the agricultural output had to be diverted to the working animals. In 1910, fully 20–25% of US farmland was devoted to feeding the animals that worked the farms.[4]

Humans directed animals for another form of mechanical energy—transportation. When we domesticated horses, we had an energy system that could produce kinetic energy to transport a person, using the energy in horse feed as input.

Before horses, transportation meant putting one foot in front of the other and walking for as long as it took to get to a destination. A human can walk about twenty miles in a day, but a horse and rider can cover twice that distance, or cover the same distance in half the time. Horse mechanical energy could also carry or pull heavier loads to transport goods or materials for building shelter.

The invention of the wheel greatly expanded the capabilities of animal-powered transportation. A wheel attached to an axle rolls over land instead of dragging through dirt and brush. By moving the load from an animal's back to a cart on wheels, the same animal could transport a much greater load. While the wheel was not an energy system per se, it significantly improved the efficiency of animal-powered transport.

Humans also employed other humans for energy. Historians tell us that the transition to agriculture enabled the differentiation of roles: Some people worked in the fields and others did not. Some accumulated wealth and were able to pay others to do hard mechanical work.

Some humans used force to direct the muscle energy of other humans. Slaves were viewed by a slaveholder as a source of mechanical energy, not unlike a domesticated work animal. Animals could deliver more power, but human slaves were more flexible in their tasks, making them a higher-quality form of energy. Wealthy Romans even used human-carried litters for transportation because a litter provided a more comfortable ride than wooden-wheeled wagons on stone streets.

Slavery was so effective an energy system that it persisted up through the 1800s. Before being banned, slavery grew to become a significant part of the economy in the American South. Southern slaveholders claimed they could not continue farming without slaves, an energy addiction similar to our modern addiction to oil.

Sail

Transportation on water became possible with the development of boats and rowing. Before there were smooth roads, boats enabled easier travel than over land. Compared to land, water has a smooth surface, no hills, and almost no friction at low speeds. Boats could float large quantities of heavy goods and people to a destination without great burden on people or animals.

Initially boats were used on lakes, rivers, and along coastlines, only gradually striking out over larger stretches of open water as navigational skills improved.

Early boats were human powered, first with a paddle, and later with oars. The Egyptians were the first to organize rowing crews and build sails, enabling the construction of large boats. Their boats were big enough to float the multi-ton stones used to construct the great pyramids from quarries as far as 500 miles upriver.

Figure 5. Egyptian sail

With a sail, the Egyptians captured the kinetic energy in the wind. Air may be invisible, but it is physical matter, and when viewed through an energy scope it carries kinetic energy when it moves. A sail puts the wind in harness and transfers a portion of the wind's kinetic energy to the boat.

The Egyptians also invented the sail, much to the relief of their rowing crews, no doubt. The Egyptian sail was a natural because the wind blows from north to south along the Nile. Since the river flows from south to north, the winds could push a boat upriver against the current, usually the hardest part of any river-rowing trip.

Initially the sail relieved the rowers on downwind runs. Later they learned to rotate the sail and angle the boat so they could even travel upwind in a zigzag fashion.

Other countries in the Mediterranean adapted the Egyptian sail to supplement their oared boats. Greeks and Romans built boats called triremes that bristled with three rows of oars on each side and lofted two sails. These ships connected all the countries of the Mediterranean Sea, spreading and intermixing goods, skills, and knowledge. Rome's ships reached as far as the British Isles.

Figure 6. Roman Trireme

The three rows of oars on a trireme enabled the ship to proceed even when there was no wind. Such reliable power made them potent warships.

Because ships could easily transport large quantities of grain, ancient Rome became dependent on grain imports from Egypt and other North African countries. If you view the grain as an energy source for Romans and their slaves and animals, Rome became dependent on imported energy. Military strategists had to devote significant military resources to protecting the sea lanes and ports for the grain shipments. We do the same today for oil shipments.

During the Renaissance, refinements of wind-powered ships extended the range of sea travel, first along the coast of Africa and then across the Atlantic to the New World. It was wind energy that enabled Magellan's ships to circumnavigate the Earth.

Europeans equipped with sailing ships were able to travel to all parts of the globe, exploring Africa, Asia, Australia, and North and South America. Those trips brought into contact all the civilizations of the world that had developed in isolation from each other. Much trauma ensued, as Europeans applied their superior kinetic-energy weapons to subjugate native peoples, and European kingdoms battled each other for control of territory.

Despite the initial violence, wind-powered ships eventually unified the human race. Once contact had been made, trade could develop and knowledge could be exchanged, even if it was often one-sided. Wind-powered sailing ships overcame the separation of people by long distances and by large bodies of water.

Rotational Kinetic Energy

Sails were also adapted for use on land, but not for transportation. In 9th-century Persia, some unknown inventor rigged sails on long arms to catch the wind to rotate a mechanism to grind grain. Grinding grain was hard work, and offloading it to a wind machine made someone's life easier.

When viewed through an energy scope, a windmill can be seen to convert some of the straight-line kinetic energy contained in wind into rotational kinetic energy in the rotating sail mechanism. Rotational kinetic energy is a brilliant human invention, and not something you find in nature.[5] Kinetic energy has many uses, but if it is straight-line kinetic energy, then it moves off and does not stay put. That's the kind

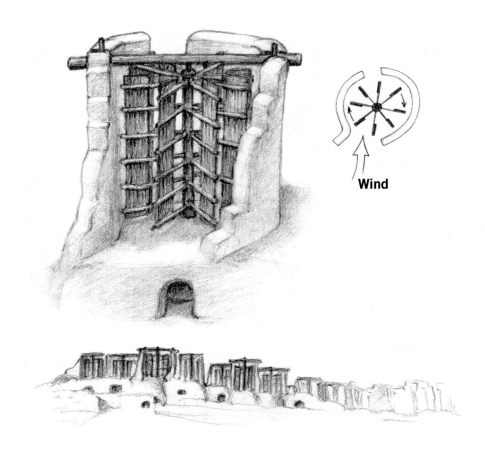

Wind

Figure 7. Early Persian windmill
By blocking one side, the wind was forced onto the other side, generating a turning motion. The millstones were in the room below. A windy location might sprout several windmills.

of kinetic energy imparted by a bow to an arrow or the wind to a sailboat. When you need kinetic energy to stay in one place to perform some stationary task, you use rotational kinetic energy. Any rotating matter has rotational kinetic energy. The matter is still moving and contains kinetic energy, but its parts are moving in circles instead of in a straight line.

Some early rotational kinetic-energy machines were animal powered. Tying a horse or an ox to a pole and having it walk in a circle created

Figure 8. Human-powered medieval winch

The worker in the wheel steps up on the wheel edge, which turns in response. By continuously stepping up, a continuous rotating motion is generated that can winch up heavy stones using an arrangement of pulleys.

slow rotational kinetic energy, which could be used to turn a stone to grind grain, or power a press for olives or sugar cane.

Medieval builders of cathedrals and castles needed to lift large stones to great heights. They winched the stones up using a windlass, with the rotational power coming from animals or humans. Medieval drawings depict squirrel-cage structures with men inside turning the cage to supply rotational kinetic energy.

Rotational kinetic energy became critically important in the industrial age. Machines to spin fibers into thread for cloth required rotational kinetic energy, initially supplied by hand or foot. The Industrial Revolution automated such processes using rotational kinetic-energy machines such as waterwheels and steam engines, followed later by electric motors and internal combustion engines. Today most electricity is produced from rotational kinetic energy turning a generator, and almost all transportation uses rotational kinetic energy, ironically to create straight-line motion of the vehicle. Many familiar mechanical kitchen appliances make use of rotational kinetic energy (blender, food processor, mixer, juicer).

But rotational kinetic energy machines go back even further than windmills and squirrel cages. As with many fundamental discoveries, the ancient Egyptians and Greeks were the first to put rotational kinetic energy to work, in the form of waterwheels.

Water Power

Before fossil fuels, the waterwheel was the most common source of rotational kinetic energy. Waterwheels could provide more power than humans or animals, and could work continuously without rest or food. The ancient Egyptians developed waterwheels for pumping water, and the ancient Greeks built the first waterwheels for milling grain:

> Spare the hand that grinds the corn, oh miller girls, and softly sleep. Deo has commanded the work of the girls to be done by the water nymphs, and now they skip lightly over the wheels, so that the shaken axles revolve with their spokes and pull round the load of the revolving stones.[6]

> —Antipatros of Thessalonica, Greece, first century B.C.E.

With waterwheels, the Egyptians and Greeks had learned to exploit a new form of energy from their environment: gravity energy.[7] Any weight elevated above its surroundings contains gravity energy. As a weight falls, its gravity energy transforms into kinetic energy, the energy of motion.

You can experience gravity energy directly when you pedal a bicycle up a hill. When you view your ride through an energy scope, you find that the extra energy you expend to push yourself uphill goes into grav-

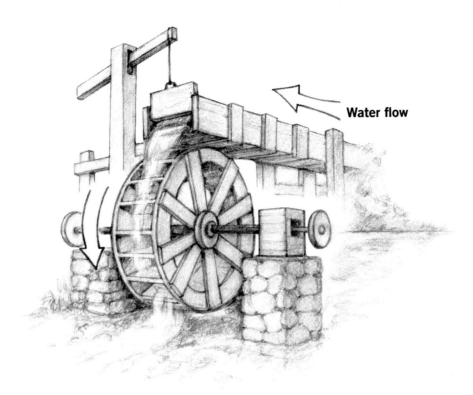

Water flow

Figure 9. Waterwheel
Pouring water onto one side unbalances the wheel and starts it turning. The rotating axle can be connected by gears or belts to turn a millstone.

ity energy stored in your body and bike. At the top of the hill, your body and bike are charged with gravity energy. How do you know? Just release your brakes. As you roll down the hill, you gain kinetic energy with no effort on your part. Your gravity energy transforms into kinetic energy as you speed downhill without pedaling.

The ancients probably made their discovery of gravity energy by observing the power of falling water. If you have ever stood under a waterfall, you know how much kinetic energy water can gain by falling a distance. Before the water falls, a waterwheel catches the water in buckets at the top of the wheel, and directs the energy of the falling water to turn the wheel.

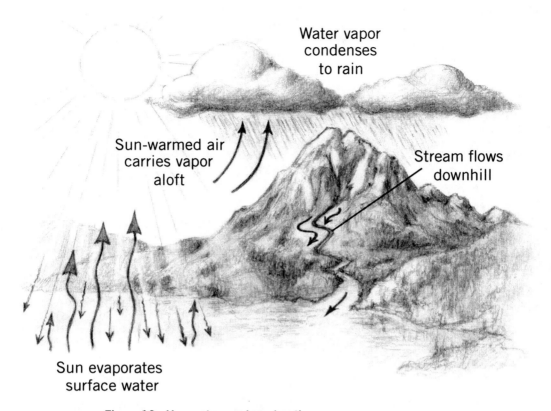

Water vapor
condenses
to rain

Sun-warmed air
carries vapor
aloft

Stream flows
downhill

Sun evaporates
surface water

Figure 10. How water reaches elevations

Solar energy hitting the surface of the oceans evaporates water into the air, where it rises on thermal air currents that are also driven by the sun's energy. When the water vapor condenses into clouds and forms rain, any water falling on land above sea level retains some of the gravity energy it received from the sun. As the water flows back to sea, the gravity energy is available to be diverted by a waterwheel for some human purpose.

A waterwheel could be connected through wooden gears or belts to drive other kinds of machinery. Building a millpond permitted gravity energy to be stored behind a dam until needed and to be finely controlled with a sluice gate (illustrated on the following page).

Over time, waterwheels were adapted for lifting mining ore and sawing wood or stone. They could also operate mechanical hammers to pound cloth into felt or work hot iron. In England, the Domesday Survey in the year 1086 counted over 6,000 waterwheels, one for every 350 people.[8]

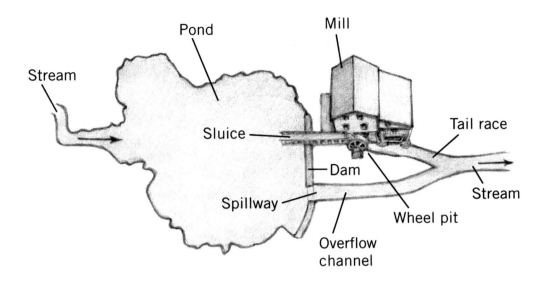

Figure 11. Millpond as energy storage

A waterwheel was often fed from a millpond, an artificial pond created by damming the stream above the mill. The miller could open a flap and draw water from the millpond whenever work needed to be done, and the pond would refill whenever natural water flow was available. The millpond served as an early (and efficient) energy-storage device, allowing the mill to operate on demand without regard to the current rate of natural water flow.

The waterwheel was the first widespread application of a stationary mechanical energy source that was not powered by humans or animals. It demonstrated how a single person could direct energy flows far beyond their own muscle power, and with far greater ease than handling animals for power.

Waterwheels helped start the Industrial Revolution, thereby opening a key new phase in the human relationship with energy. The Industrial Revolution began with the automation of cloth making. Richard Arkwright patented his *water frame* thread-spinning machine in 1768. He called it a water frame because it combined a waterwheel for power and a frame to handle the spinning spindle. Another machine of the time, the *spinning jenny*, could handle eight or more spindles at once, but it was powered by the spinner's right hand. Eventually the two were combined to produce machines capable of handling hundreds of spindles, all powered by a waterwheel.

Before the development of the steam engine, the textile industry was heavily dependent on waterwheels. The waterwheel was so important that the site of a textile factory depended on its having access to water power.

Gunpowder

During the Wood Energy Epoch, human weapons transitioned to wood energy as well, in the form of gunpowder. Before gunpowder, our weapons had always been powered by muscle energy. The discovery of gunpowder in 9th-century China introduced chemical energy into weapons, revolutionizing hunting and warfare.

With a spark to ignite it, the powder charge in a gun burns quickly to transform the stored chemical energy into heat energy, while forming carbon dioxide and nitrogen gases as combustion products. The trick is to get the mixture right so it forms heat quickly, but not so quickly that it creates a shock wave that blows your gun apart. The heat energy causes the gases to expand, and if the mouth of the gun barrel is the bullet's only way out, then the expanding gases propel the bullet down the barrel at great speed. About a third of the heat energy is converted to kinetic energy of the bullet, the rest ending up in hot gases and a heated barrel.[9]

Of course, the efficiency of energy conversion matters little when the efficiency of killing increases so much. Steel armor could deflect arrows propelled by human muscle power, but bullets propelled by gunpowder could penetrate armor, making armor obsolete. The cannon, a scaled-up version of the gun, made castle walls obsolete.[10]

A military viewed through an energy scope can be seen as a large complex energy system designed to shower destructive kinetic energy on a designated enemy. At all levels, from small weapons fire through cannons and missiles, all can be viewed as energy devices delivering kinetic energy in excessive quantities onto enemy humans to inflict damage on them. Despite all the benefits we derive from energy, it has this dark side as well.

The original gunpowder was made up of charcoal, saltpeter, and sulfur. The charcoal, which provides most of the energy, comes from trees whose energy derives from photosynthesis. So gunpowder can be seen as another form of solar energy.

PART I: HOW WE GOT HERE

CHAPTER 4

The Fossil Fuel Epoch

J ust as gunpowder shaped military history, so did fossil fuels shape industrial history. The abundance of fossil fuel energy made possible the rapid expansion of industry and automation. It was the invention of the steam engine that initiated the transition from the Wood Energy Epoch to the Fossil Fuel Epoch.

Steam Engine

In a surprising turn of events, it was gravity that led to the development of steam engines. The first steam engines were not built to power textile factories but to counteract gravity and lift water out of coal mines.

People had burned coal in England as early as the Bronze Age, as evidenced by the coal ash found in some funeral pyres dating from 3,000 to 2,000 BC in Wales.[1] Surface outcroppings of this burnable rock provided easy access to a fuel for heating and cooking for hundreds of years. The British pioneered the first fossil-fuel commodity industry when they regularly transported and sold coal in London. Coal burning in medieval times was so prevalent there that its smoke created the first air pollution from fossil fuels, prompting King Edward I to proclaim a ban on coal burning in London in 1272 (with few other options for energy, most Londoners were forced to ignore the ban).

By the 13th century, most of the easily harvested surface coal in England had been consumed, so the first mine shafts were dug into coal seams. There, miners experienced one of the most vexing problems of mines: water seepage.

A mine shaft extending below the underground water table will fill up with water that seeps in through the walls. To work the mine, the water has to be lifted out of the mine as fast as it seeps in.

When viewed through an energy scope, this water-lifting operation was the opposite of a waterwheel. Just as you can extract energy from falling water, so you must invest energy to lift water. The sun lifts water naturally to power waterwheels, but the sun does not reach the bottom of a coal mine, so some other energy source must be employed. In the early mines, teams of men or horses had to continuously work buckets or pumps to keep the mine from flooding, making the coal more expensive to extract.

Human-powered coal extraction

Before steam engines, all coal-mining work required muscle-straining human labor. The hewer was concerned solely with cutting the coal. It was the business of the barrowman or putter to fill the coal into corves or baskets, load these on the wooden sledges or trams, and drag, push, or haul them to the pit bottom. At the pit bottom an onsetter hung the corves on the rope, and a brakesman or windsman drew them to the pit-eye and delivered them to the two banksmen, who carried them on a sledge to the coal heap.[2]

Necessity being the mother of invention, various tinkerers tried to work out how to lift the water using fire energy from burning some of the coal available at the mine. None were fully successful until 1712, when Thomas Newcomen installed the first practical mechanical steam engine to pump water out of an English coal mine. (See the following illustration.) By the time of his death seventeen years later, over a hundred of his engines had been installed.

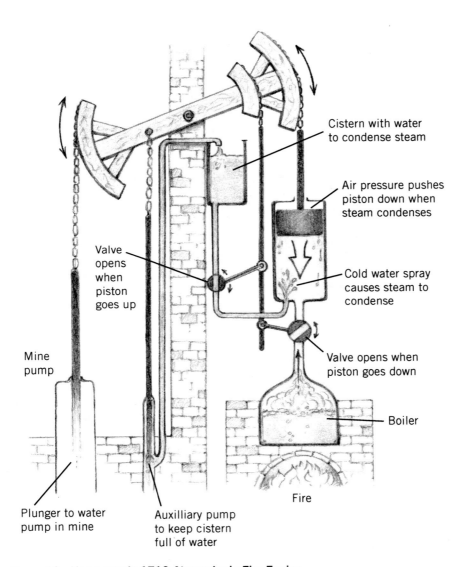

Cistern with water
to condense steam

Air pressure pushes
piston down when
steam condenses

Valve
opens
when
piston
goes up

Cold water spray
causes steam to
condense

Mine
pump

Valve opens when
piston goes down

Boiler

Fire

Plunger to water
pump in mine

Auxilliary pump
to keep cistern
full of water

Figure 12. Newcomen's 1712 Atmospheric Fire Engine
Newcomen's machine condensed steam back to water to form a vacuum, which sucked on a piston to drive a pump. Although Newcomen's steam engine was less than 1% efficient, it replaced the manual labor of many men and horses, and was economical as long as it liberated more coal than it consumed.

Newcomen's steam-powered water pump was the first of a class of machines called *heat engines*. A heat engine is any device that converts

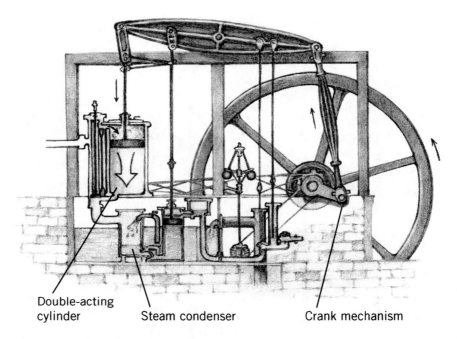

Double-acting
cylinder   Steam condenser   Crank mechanism

Figure 13. Watt's 1784 Steam Engine

Watt's engines operated on the same principle as Newcomen's, by creating a vacuum to suck on the piston. His contributions were a double-acting cylinder which could produce power in both directions, and a separate steam condenser to keep the power cylinder hot. He also adapted a crank system to convert the up-and-down motion of the piston into rotary motion.

heat into mechanical motion. Typically, fuel is burned to make the heat, so any motor that burns fuel is a heat engine. The class of heat engines grew over time to include modern steam turbines for generating electricity, gasoline and diesel engines that power cars and trucks, and jet engines that lift airplanes. Heat engines extended fire into new realms of human energy use.

No one helped the cause of steam engines more than James Watt. His innovations enabled steam engines to break free of their initial role as simple water pumps to become universal power sources. Watt, whose name is often associated with the steam engine, did not invent it; rather, his contributions greatly improved their efficiency and adapted them to many new applications, especially rotational ones.

Once a steam engine could produce rotational kinetic energy, it could perform many more mining tasks than just pumping water. For example, a steam engine connected to a winch could lift the baskets of coal up out of a mine. Over time, almost all manual tasks in mining were replaced by equipment powered by heat engines.

Figure 14. Factory powered by single steam engine (Germany 1905)
A Watt steam engine could power an entire factory using the versatility of rotational kinetic energy. The single rotating power shaft from the steam engine could be linked to other rotating shafts with belts and pulleys. By engaging and disengaging a belt, power could be turned on and off to a particular machine. By arranging different pulley diameters, the rotational speed could be stepped up or down. Any machine designed to use rotational kinetic energy could thus be powered by a central steam engine. A factory ceiling became a maze of moving shafts, pulleys, and belts.

The rotary power of Watt's engines enabled them to move into other industries beyond mining. The Albion Mill in London installed two Watt engines in 1783, each capable of powering ten sets of millstones to grind wheat into flour.[3] Just as the growing textile industry was running out of suitable sites for waterwheels, the Watt steam engine became available to replace water power. Ironically, in some cases during the transition period, a steam engine was deployed to drive a pump to lift water above an existing waterwheel that was already set up to power the textile machinery. It soon became clear, however, that a steam engine could run the machinery directly.

Watt's steam engines were huge stationary machines, usually requiring a building of their own. Their great expense, and the risk associated with this new invention, caused many potential customers to hesitate. In most cases Watt overcame their resistance through his novel marketing scheme. He did not sell his engines, but installed them for free in exchange for one-third of the value of the energy saved from replacing horses, humans, or even Newcomen's less efficient engine.

Since his potential customers required proof of his claims, Watt had to develop accurate methods of comparing energy flows.

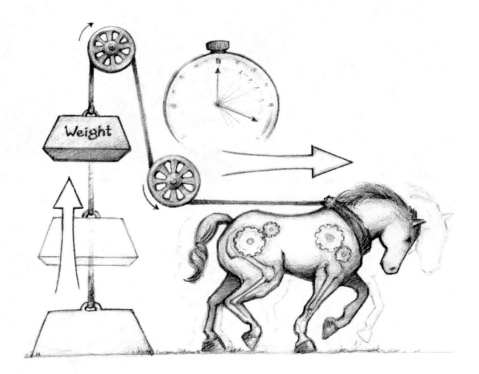

Figure 15. Measuring Horse Power

Watt harnessed a horse to a pulley to lift a carefully measured weight to a given height. He then compared that effort to the weight lifted by a winch powered by his steam engine.

Watt found that the key measurements were not only the weight and the height, but also how quickly the weight could be lifted to that height. The rate at which energy could be delivered ultimately determined the rate of production of goods that a factory could achieve when powered by his steam engine.

A rate of energy flow is called *power*. Watt measured the power of his horses and engines in units of *foot-pounds per second*, with the lift height measured in feet, the weight measured in pounds, and the time measured in seconds. For his customers, he converted this rather technical sounding term into *horses equivalent*, so he could tell his customers how many horses his engine could replace. Since the customer generally knew how much a horse cost to operate, they could determine the value of the power being supplied. This measure later came to be called *horsepower*, which today is standardized as 550 foot-pounds per second.

What is power?

There are many meanings of the word *power*, such as political power or military power. Here we are describing physical power, which is precisely defined as the rate at which energy flows. Although the words power and energy are often used interchangeably, they differ in a crucial way: Power is energy in motion. Fossil-fuel energy is a static commodity that does nothing while sitting in the ground for millions of years. Only when the fuel is burned can it generate power to accomplish some task.

Even with his improvements, Watt's steam engines were huge machines with low efficiencies, which limited where they could be used. The deciding breakthrough came in 1799 when Richard Trevithick created the first high-pressure steam engine. Instead of using steam at atmospheric pressure to create a vacuum, Trevethick closed the system with strong, tight fittings that let the steam build up pressure. He passed that pressurized steam to the cylinder to push on the piston instead of pulling it. With higher pressure, he could produce higher power, and do that in a smaller apparatus.

The smaller sizes of pressurized steam engines made it possible to use them in transportation. In 1825 the first public steam-powered railway began operating in England. Steam locomotives were quickly deployed in other countries, and a network of steam-powered rail transportation grew up. Railroads revolutionized land transportation by carrying hundreds of passengers and tons of goods in a single train, and at speeds unheard of on the dirt roads of the time.

Steam engines were quickly adapted to ships as well, initially for river and lake transportation, and later as ocean-crossing ships. Their great advantage over sailing ships was that they were not dependent on the wind, an energy source that could be highly variable in different locations. The disadvantage was that they had to load and carry all the fuel they needed to reach a destination—and enough to get back if there was no fuel supply at the destination.

The smaller pressurized steam engines also found uses in agriculture. One application was as a portable power source for threshing grain. The steam engine was transported to a farm at harvest time, and set up in a central location to process all the harvested grain brought to it. Steam engines were also outfitted with wheels powered by the engine, forming the first tractor that could replace a team of horses for plowing or towing harvesting equipment.

Development of the steam engine was a pivotal event in the human relationship with energy. In a sense, it was yet another human application of fire, but it was the first time fire could be used for mechanical energy. And unlike wind and water power, a steam engine was portable and could be set up anywhere. Textile factories no longer needed to be built next to a source of water power.

The advantage of a steam engine is obvious: A single steam engine could do the work of many men and many horses. A single person operating a loom powered by steam could do work equivalent to dozens of hand weavers. A steam-powered sawmill could relieve whole crews of workers laboring to pull hand saws. The steam shovel invented in 1839 could replace armies of human diggers in mining and building operations.

People refer to "steam power" without realizing that steam is not the source of energy, just part of the mechanism to transform the energy from heat to kinetic energy. The source of energy is always the fuel fed into the steam engine.

Early steam engines were powered by wood fuel harvested from local forests. The voracious appetites of those inefficient engines, however, quickly stripped the nearby forests. Coal, which had been used as a source of heat energy for centuries in England, became the primary heat source for steam engines. After 1830 most locomotives were coal-fired.

The transition from wood to coal for steam engines set us on the path that eventually led to our current dependence on fossil fuels.

Like most early energy inventions, the first steam engines were not built by academics based on scientific principles, but by tinkerers seeking to solve a practical problem.

That heat could produce mechanical motion in a steam engine to lift water out of mines fascinated the scientists of the early 1800s. The nature of heat was a mystery then. You cannot see heat, but clearly it flows from hot to cold, as you know if you have ever grabbed the handle of a hot cast-iron frying pan.

Since heat flows, many thought heat must be an invisible fluid, which they called *caloric fluid*. But when they tried to determine the nature of caloric fluid by, say, weighing it, they discovered that a hot body does not weigh more than a cold one, so caloric fluid had no weight.

The mystery of heat deepened in 1798 when Benjamin Thompson, also known as Count Rumford, published a paper describing a curious experiment. While supervising the boring of cannon barrels in Munich, he had noticed that the barrels became very hot. For his experiment, he immersed a cannon barrel in water. After several hours of boring with a dull bit, the water boiled without any fire being applied. He had demonstrated that mechanical motion could produce heat through friction (you can rub your hands together vigorously to demonstrate that fact).

Friction is something everyone is familiar with. While riding a bicycle on level ground, if you cease pedaling, you will gradually coast to a stop. You experience friction in the wheel bearings rolling in the hub, friction in the tires rubbing the ground, and friction in the air passing around you. Friction gradually converts your kinetic energy into heat energy. That's generally what friction does, convert motion into heat. The heat is spread out so thinly that you don't feel any rise in temperature. Although the mechanical effects of friction cannot be ignored, the heat it generates generally is. In Thompson's experiment, the heat of friction became dramatically obvious when it boiled the water.

What is heat?

At that time they could not know that heat was actually just random kinetic energy at the molecular level, too small to be seen. If you viewed your warm body through an energy scope, you could see every molecule of your body humming with vibrational kinetic energy. Heat flows from hot to cold by passing along those vibrations.

James Joule's Discoveries

Steam engines showed that heat could be transformed into kinetic energy to lift water, and Thompson showed that kinetic energy could be transformed into heat to boil water. That the energy could flow in either direction seemed to indicate that heat and kinetic motion were two forms of the same thing: energy.

James Joule (rhymes with *tool*), an English brewer turned self-funded scientist, provided the most convincing evidence of this link in 1845 at the age of twenty-seven. Rather than ignoring friction, Joule's experiments centered on the heat of friction. Joule's toy-like apparatus that unified the understanding of energy is illustrated on the next page.

Energy Equivalence

This principle states that any form of energy can be converted to any other form of energy. A light bulb converts electricity to light energy, while a solar panel does the reverse and converts light energy to electricity. Although some transformations are difficult, inefficient, or impractical, none are impossible.

Joule's experiments signaled two real breakthroughs for the science of energy. First, they showed that diverse forms of energy such as gravity energy, kinetic energy, and heat, which appear to be not at all similar, could be transformed one into another, indicating a deep connection among them. He had established the *equivalence of energy*, a powerful notion that Einstein later used to develop his Theory of Relativity.

In Joule's second breakthrough, he showed that during energy transformations, the total quantity of energy was preserved. The amount of heat energy that was generated by the turning paddles exactly matched the amount of gravity energy lost by the descending weight. Something was being preserved, even as it changed its form. That something was energy.

Since then, experiments by other scientists using many forms of energy in many different energy transformations have always come to the same conclusion: Energy never disappears. These experiments established what is now called the *Principle of Conservation of Energy*, which says that while energy may change forms during a process, the total amount of energy remains the same; the energy is conserved.

This principle is often restated in a more familiar form: Energy can neither be created nor destroyed.

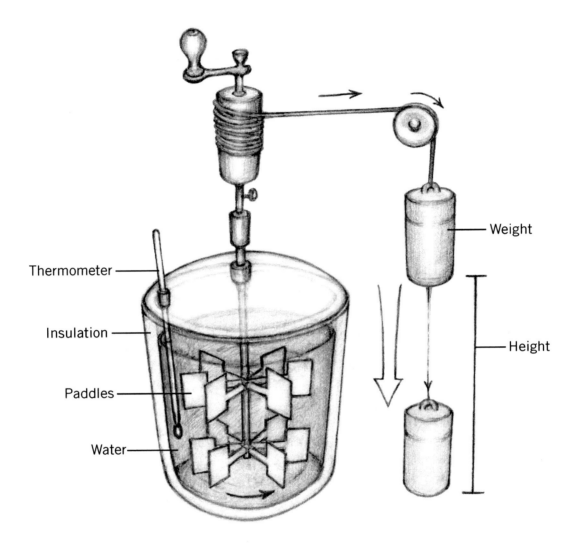

Figure 16. James Joule's energy experiment

Joule hung a weight from a string, and wrapped the string around an axle attached to paddles suspended in water. When the weight descended, the paddles turned and stirred the water. The friction from the stirring generated heat that raised the temperature of the water. By using highly accurate thermometers, and by carefully accounting for any energy leakages such as pulley friction, Joule was able to compare the gravity energy before the descent to the heat energy generated by the descent, and determined they were the same amount of energy.

The implications of this principle are astonishing. The second half of the statement, that energy cannot be destroyed, essentially says that *energy is indestructible.* That seems counterintuitive because when you fill your car's tank with gasoline energy and drive the car, the gasoline burns up and disappears. But the gasoline is just the material carrier of the energy, which is stored in chemical bonds in the gasoline. As you drive the car, the gasoline burns in the engine, breaking those chemical bonds and transforming the chemical energy into heat energy, while destroying the carrier.

"Energy can neither be created nor destroyed"

The heat energy remains, powering the car and escaping out the exhaust pipe. If you account for all the paths of energy transformation that follow, you find that while the storage medium of the energy is destroyed, the energy itself is not.

Energy's property of indestructibility has proven to be absolute. While humans can manufacture objects that can survive harsh conditions, they eventually wear out and break down. But energy is absolutely and totally indestructible. There is not a single known case where energy was destroyed and disappeared from our universe.[4] Nothing else in our physical universe has that property, making energy unique.

So an energy process is best described as a *flow* of energy from one form to another. In any energy flow, energy changes form, *trans-forming* from one kind of energy to another. The old form disappears and the new form appears in its place. Every event in our universe involves some type of energy flow.

Putting aside philosophical abstractions, energy science largely becomes an accounting process. Just count the energy going into an energy flow, and count the energy coming out of it. That the energy coming out has a different form from the energy going in does not change the hard requirement that the books must balance.

This conclusion carries huge implications for human energy use. According to this rule, any energy we want to use must come from somewhere. When we have a task to perform that requires energy, we must line up a source of energy to power that task. We cannot print new energy, so to speak.

That might sound like a limitation that would inhibit human activities, but in fact it greatly expanded our ability to use energy. With the rules of energy accounting firmly, accurately, and irrevocably established, humans proved capable of predicting the future. That is, by setting up an energy transformation process and supplying it with measured amounts of certain forms of energy, scientists and engineers can describe how the energy transformation will proceed and what the outcome will be, before it takes place. That effectively amounts to predicting the future.

For example, by loading up a rocket carrying a moon lander with a certain amount of chemical energy in the form of rocket fuel and pointing it in the right direction, engineers can predict that the rocket will reach the moon. By accounting for the chemical energy in the fuel, the weight of the rocket, and the efficiency with which the fuel is converted to kinetic energy, engineers can predict the precise path of the rocket, how long it will take, and whether it will reach its fast-moving target a quarter-million miles away. If, en route, some energy was mysteriously added or removed, then the rocket would veer unpredictably off course. The Principle of Conservation of Energy proved sufficiently reliable that we could risk human lives in sending six sets of astronauts to the moon and back in the Apollo space program.

That kind of magic makes the science of energy supremely potent. The discovery of the Principle of Conservation of Energy permitted the human race to safely control ever-larger amounts of energy with ever-finer degrees of precision. Much of our modern technology is based on that energy precision.

James Joule's simple experiments had such profound practicality that they proved to be a turning point in the human relationship with energy. Before Joule, we learned to control energy through trial and error, without really understanding what we were doing. After Joule, we used the Principle of Conservation of Energy to tease out all the other rules of energy. That knowledge enabled us to design new energy systems on paper and predict how they would work, establishing the human race as the masters of energy flows.

James Joule's work was so important that in 1889, the year of his death, the standard metric unit of energy, the *joule*, was named in honor of him.

That honor was appropriate, because the need for a new unit of energy was a direct result of Joule's work. Before Joule, each form of energy was studied separately, and was measured in terms of its own characteristics. As described earlier, Watt measured gravity energy as pounds lifted to a given height, in units of *foot-pounds*. The unit of measure for heat energy was the *calorie*, defined as the amount of energy needed to raise the temperature of one gram of water by one degree Celsius.[5]

That mechanical energy and heat could be derived one from the other showed that they were two aspects of the same thing: Energy. Energy equivalence suggested the need for a common unit of energy measurement, not unlike the introduction of the euro as the common unit of currency in Europe.

The joule unit of energy can be used to measure any and all forms of energy. That's helpful when comparing measurements of different forms of energy, as you must do when tracking energy transformations.

How much energy is a joule? Not much. A joule is equivalent to about three-quarters of a foot-pound, so you could say a joule is the energy needed to raise a three-quarter-pound book by a foot, such as lifting a book to the next higher shelf.

A joule is not a large unit of energy compared to what modern people use. One gallon of gasoline contains 132 million joules of chemical energy, so you could easily cart around a billion joules in your gas tank. A single gasoline joule would cost you about three microcents (three millionths of a cent).

Large energy numbers bring into play the multipliers that the joule unit shares with other quantities. In computers, a megabyte (MB) of memory is a million bytes, and in energy, a megajoule (MJ) is a million joules. So you can say a gallon of gasoline contains 132 megajoules of energy. These multipliers make it easy to grasp large numbers without having to count the zeros. Here is the whole set of available multipliers:

- kilojoule (KJ) = thousand joules (10^3 joules). One AA battery stores about 10 kilojoules. The nutrition labels on food sold in the European Union often express food energy in terms of kilojoules.
- megajoule (MJ) = million joules (10^6 joules). The food energy in one big slice of chocolate cake is about 1 megajoule.
- gigajoule (GJ) = billion joules (10^9 joules). An eight-gallon gasoline fill-up contains about 1 gigajoule.
- terajoule (TJ) = trillion joules (10^{12} joules). An underground gasoline storage tank at a filling station holds about 1 terajoule.

- petajoule (PJ) = thousand trillion joules (10^{15} joules) A small oil tanker carries about 1 petajoule, and the largest supertankers hold up to 24 petajoules.
- exajoule (EJ) = million trillion joules (10^{18} joules). Total US energy use in 2009 was 100 exajoules.
- zettajoule (ZJ) = billion trillion joules (10^{21} joules). Total energy in known fossil fuel reserves is about 40 zettajoules.
- yottajoule (YJ) = trillion trillion joules (10^{24} joules). The sun puts out 383 yottajoules each second.

As you can see, this system provides a very compact way to express some really big numbers, which we will need when discussing patterns of world energy supply and demand. For example, you can express the total energy contained in all the coal, oil, and natural gas reserves in the world as simply "40 ZJ." Or you could express it the long way as "1,340 billion barrels of oil, 6,261 trillion cubic feet of natural gas, and 909,393 million short tons of coal."[6]

Once the joule was established as the standard unit of energy, it could also be used to describe power, the speed of an energy flow. Just as a speed of travel can be measured in miles per hour, so can a speed of energy flow be measured in joules per second.

A power measurement captures the essential story of any energy flow —how *much* and how *fast*. Power describes energy in action.

This basic unit of power—joules per second—is used so often that it has been combined into its own unit, the *watt*, appropriately named after James Watt who emphasized the need for measuring power.

One watt is equivalent to one joule per second. A 25-watt light bulb consumes electricity at the rate of 25 joules per second. In our travel analogy, the watt is similar to the nautical knot, a single unit that expresses a rate of travel at sea. Note that you would never say, "watts per second," just as you would never say "knots per hour," because time is implicit to the unit.[7]

The watt shares the same set of multipliers as joules for larger quantities, so you will see kilowatt, megawatt, gigawatt, and terawatt used in this book. And who can resist using yottawatt in a sentence, as in "The sun continuously puts out 383 yottawatts of power."

High Explosives

As we gained better understanding and control of energy in the 1800s, we stepped up the amount of energy we could safely control. With the development of high explosives such as nitroglycerin and dynamite, we began to reshape the surface of the planet.

You might not think of an explosive as an energy system, but a view through an energy scope can give you a different perspective. In any energy system, humans direct energy flows for some human purpose. When road engineers place explosives in specially placed holes, they are managing a controlled release of chemical energy to break apart the rock so it can be excavated. An explosive may be dangerous, but when properly controlled, its sudden energy release can serve our purposes.

And outside of military applications, that purpose is generally reshaping the landscape. Earth-moving equipment cannot move solid rock. Dynamite and other explosives must first blast rock loose to enable mining, road building, railroad building, dam building, and canal building to proceed.

Explosives have been used since the Chinese invented black powder in the 9th century. Black powder is considered a low explosive, because the speed at which it burns is less than the speed of sound. The burning process generates hot gases, which if contained will rapidly build up pressure until they explode their container. When black powder is used in a firearm, the pressure never reaches that level because the bullet in the barrel moves out, relieving the pressure. If a bullet jams in the barrel, the gun will explode.

In 1847, Italian chemist Ascanio Sobrero invented nitroglycerin, the first high explosive. The burn rate of a high explosive exceeds the speed of sound, creating a supersonic shock wave that amplifies the force. Unfortunately, nitroglycerin was so unstable that its inventor argued vigorously against its use as an explosive.[8] A physical bump is enough to detonate nitroglycerin, and as it ages, it degrades into even more unstable forms. Transporting nitroglycerin was extremely hazardous on the rough roads of the time.

Swedish chemist Alfred Nobel (whose bequest established the Nobel Prizes) stabilized nitroglycerin in 1867 by mixing it with diatomaceous earth to form *dynamite*. Nobel was no doubt motivated by the death of his younger brother, Emil, in 1864 from a nitroglycerin explosion at their father's factory.

Nobel derived the name from *dynami*, the Greek word for power. And dynamite did grant us unprecedented power over the land. Dynamite played a crucial role in excavating the Suez Canal in 1869 and the Canal at Corinth in Greece in 1893, a project attempted unsuccessfully in Nero's day using slave labor. The fifty-one-mile-long Panama Canal would not have been feasible without dynamite.

Dynamite was used for blasting the tunnels for the London Underground and for New York's subways. Most large dams for hydroelectric energy production required dynamite to shape the land before building the dam. Dynamite also supported the construction of many of the world's great highway systems.

In the 20th century, plastic explosives such as C-4 have improved the control, and hence the safety, of explosives for civilian use. Today demolition experts using precisely timed sequential explosions at key support points can neatly drop a building in seconds.

In a nearly forgotten chapter of US nuclear history, Operation Plowshare flirted with using *nuclear* explosives for excavation. The 1962 Sedan nuclear test shot in Nevada blasted a crater 1,280 feet across. It also spread radioactive fallout as far as Iowa and South Dakota, exposing more Americans to radioactivity than any other nuclear test. For that reason, the idea was abandoned.

Even with just conventional explosives, humans seem to have no barriers to reshaping the land. If a mountain is in our way, we remove it. Only humans control enough energy to accomplish that.

Electricity

In terms of sheer usefulness to the average person, the most important energy advance of the 19th century was the introduction of electricity service.

In 1820, Hans Christian Ørsted of Denmark noticed that his compass needle moved when he ran an electric current through a wire. He had discovered that an electric current generates its own magnetic field—and that field affected his compass. One year later, British scientist Michael Faraday replaced the compass needle with a rotating magnet to demonstrate rotary motion derived from electricity. By creating a device that transformed electricity into kinetic energy, he had invented the first electric motor.

Ten years later, Faraday discovered the opposite effect—by moving a magnet near a wire, an electrical current could be generated. With further refinement, Faraday invented the electrical generator, a device that transforms kinetic energy into electricity.

Once Joule's work showed how to account for energy inputs and outputs, engineers quickly developed practical and efficient electrical generators and motors. The combination of the two introduced the idea of "energy at a distance," or energy transmission. If the shaft of a coal-fired steam engine were connected to an electrical generator, the rotational kinetic energy from the steam engine could be transformed to electricity. That electricity could then travel over a wire to a distant electrical motor, where it would be converted back to rotational kinetic energy to drive a machine. That machine was effectively driven by a coal-fired steam engine located many miles away, using electricity as the intermediate form of energy transmitted over that distance.

This combination quickly broadened the human control of energy through electric utility service, introduced by Thomas Edison in 1882. Edison connected a coal-fired steam engine to an electrical generator to supply electricity to fifty-nine customers in a section of New York City.

Edison built the plant not so much because he wanted to start the first electric utility company, but because he wanted people to be able to operate the electric light bulb he had invented. Only three years elapsed between the invention of the bulb and the start of electricity service for it. Electric lights quickly established themselves as a superior form of lighting because they were brighter, safer, and easier to turn on and off. Once electricity became available for lighting, other uses for electricity service quickly followed. Electric motors proved to be versatile and reliable suppliers of rotational kinetic energy, soon replacing steam engines in factories.

If you need kinetic energy to push, pull, lift, turn, pump, stir, or vibrate anything, it is hard to beat an electric motor. They are quiet and efficient, and can be used in enclosed spaces because they emit no fumes. Most electric motors have a single moving part, the rotor, and so require almost no maintenance. They come in all sizes, from the microscopic single-molecule nano motor created at Tufts University[9] all the way up to the giant 480-million-watt motor at the Bath County Pumped Storage Station in Virginia.

Rising electricity demand soon showed that the old piston steam engines were not well matched to electrical generators because they turned too slowly. Charles Parsons patented the steam turbine in 1884 to meet that need.

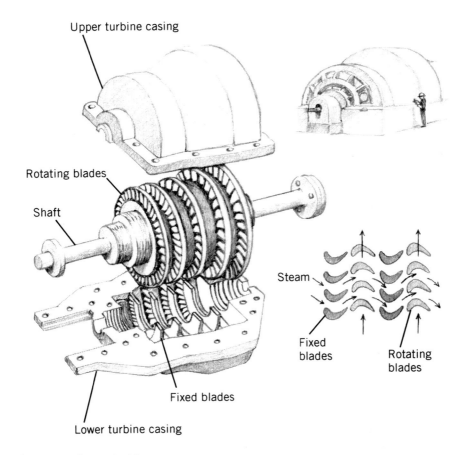

Upper turbine casing

Rotating blades

Shaft

Steam

Fixed blades

Rotating blades

Fixed blades

Lower turbine casing

Figure 17. Steam turbine

The fixed blades in the steam turbine direct the steam against the moving blades, thereby keeping the flow smooth and allowing multiple stages to convert the most energy. A typical piston steam engine turned its shaft at about 150 rpm (revolutions per minute), while a steam turbine typically operates at 3,000 rpm.

In a steam turbine, fast-moving steam pushes fan blades that rotate at high speeds. The shaft of the turbine can be directly coupled to an electrical generator. In just twenty years steam turbines had completely replaced piston steam engines for generating electricity.

Even in its early days, electricity made possible entirely new applications of energy. Rotational kinetic energy no longer needed to be tied to a steam engine or waterwheel, but could be supplied by an individual electric motor scaled to the need. It is unlikely we would have ever seen a steam-powered vacuum cleaner, but electric vacuum cleaners appeared in the early 1900s, shortly after electricity became available to businesses and households.

Electricity is the most versatile form of energy because it can be so easily converted to almost any other form of energy. In addition to kinetic energy from a motor, you can produce heat energy from a heating coil, light energy from a light bulb, and sound energy from a speaker. Electricity can be converted to chemical energy for storage in a battery, or to gravity energy by pumping water uphill.

Electricity also revolutionized communications by enabling instant communication at a distance. First the telegraph and later the telephone could send coded electrical impulses over a wire for almost any distance. Their value was not in the energy being delivered but in the information carried by the energy. Only tiny amounts of electricity were needed, yet they had great impact on the development of human society by breaking down the barrier of distance. Today's Internet and smart phones emerged as their direct descendants, providing exquisite control of minute amounts of electrical energy for computing, communication, and entertainment.

From a consumer's point of view, electricity is a source of energy. You need only insert a plug and you have power on demand. But there are no natural sources of usable electricity,[10] so it must always be generated by transforming energy from some other form.

Hydroelectric power plants have converted gravity energy to electricity for decades, starting at the same time as the earliest steam-powered plants. But hydroelectric plants could only serve cities near a source of elevated water. Heat engines fed with coal, and later oil or natural gas had no such constraints, and so grew to fill the growing need for electricity. Today, two-thirds of electricity worldwide is produced using heat engines burning fossil fuels.[11]

Electricity demand has kept coal in demand. Currently coal accounts for 42 percent of world electricity production, the largest single source of electricity. Coal plants are still being built in countries such as China and India where electricity demand is growing.

The one area where coal and electricity had trouble competing was in transportation.

Internal Combustion

In the days before automobiles, personal transportation was powered either by your own muscles on a bicycle, or by a horse.

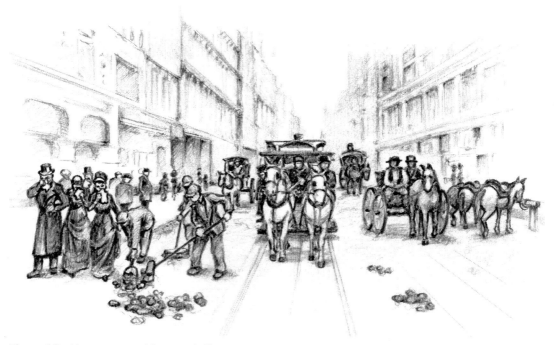

Figure 18. Horse-powered transportation
Cities had become crowded with personal horses, private coaches, cabs, public buses, and delivery vehicles, all horse-drawn and all creating sanitation problems from the manure.

By the late 1800s, many tinkerers were trying to replace the horse with an electric motor or steam engine so they could sell a "horseless carriage."

Some of the first automobiles were electric. In fact, the first car built by Ferdinand Porsche, founder of Porsche AG, was electric powered. This 1898 electric car resides in the Porsche Museum in Stuttgart, Germany.

An electric motor was easy to install, easy to start, and easy to drive because it did not need a gearbox. But the electricity had to come from on-board batteries, and the batteries of the day were heavy, weak, and prone to failure (and expensive to replace).

By the late 1800s, pressurized steam engines could be made small enough to fit into an automobile. Like electric cars, steam cars did not require a gearbox and were easy to start, though they needed a few minutes to initially build up steam. Unlike electric cars, steam cars could travel long distances at high speeds. Successful models included the *Runabout* from the Locomobile Company of America and the *Stanley Steamer* from the Stanley Motor Carriage Company.

The late 1800s also saw the development of a second type of heat engine, the *internal combustion engine*. Nicholas Otto patented the four-stroke engine in 1886 (illustrated on the following page), and Rudolf Diesel patented his compression-ignition design in 1895. These two basic designs are still used today in most cars and trucks.

Internal combustion engines are so named because they burn a liquid or gas fuel directly inside a cylinder rather than in an external boiler as with a steam engine. When the fuel burns, the hot gases create pressure to push the piston, which ultimately pushes the car.

The combustion gases, not steam, are the working fluid, and the working fluid is thrown away at the end of each cycle, flowing out the exhaust pipe along with two-thirds of the heat energy. Liquid fuels derived from oil were the preferred energy source because they could easily be pumped into the car and stored in a tank.

Oil from natural oil seepages had been used for centuries in lamps for lighting. Called *rock oil* (petr-oleum), such natural seepages were the only source of crude oil until successful wells were dug starting in 1859.

At about that time, Polish pharmacist Ignacy Łukasiewicz created the process to extract clear kerosene from crude oil. On July 31, 1853, Łuka-

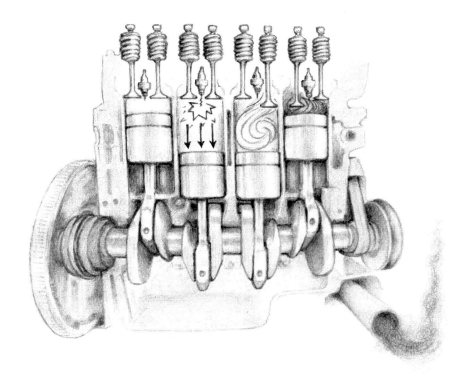

Figure 19. Internal combustion
The first piston downstroke draws in a mix of air and fuel through the intake port. The first upstroke compresses the mixture and then ignites it. The second downstroke is pushed by the expanding hot gases in the cylinder, providing the power. The second upstroke pushes the spent gases out through the exhaust port. Amazingly, this process cycles ten or more times per second in each cylinder, enabling the engine to build significant power.[12]

siewicz provided one of his kerosene lamps for a surgical operation, and it performed so well that the hospital bought several. On that day, the modern oil industry began, say historians. Łukasiewicz went on to start the first refinery and later the Polish National Petroleum Society.

Kerosene quickly replaced the more expensive whale oil in lamps, leading to the decline of the whaling industry. Łukasiewicz and the nascent oil industry probably saved whales from extinction.

Once liquid fuels could be extracted from crude oil, the development of the internal combustion engine proceeded rapidly. The main advantage of an internal combustion engine over a steam engine is its size. Without the need for a boiler and condenser, an internal combustion engine can be made smaller and lighter than a steam engine of equivalent power. Today you can hold a gas engine in your hand, in the form of a leaf blower or chainsaw.

But getting the early car engines to start was difficult, because a piston had to be moving to suck in the fuel mixture. The car operator had to manually turn the engine with a hand crank until the cycle could sustain itself. Turning the crank required considerable strength, and a backfire could flip the crank with enough force to break an arm. This single difficulty temporarily left space for steam and electric cars to compete. That space contracted in 1911 when the Cadillac Motor Company attached an electric starter motor and battery to a gasoline engine to eliminate the hand crank. Ironically, an electric motor killed the early electric car.

Mass production on Henry Ford's conveyor-belt assembly lines also made gasoline-powered cars more affordable than steam cars. The last Stanley Steamer was built in 1924, and by 1927 over 15 million Model T Fords had been sold.

Soon anyone could own a personal heat engine. This was not a hulking steam engine needing a dedicated building but a compact device that could be stored in a garage. It enabled an individual to generate motive power on demand. With this heat engine, you could not only travel fast like a train, you had the freedom to travel when and where you chose.

That is, as long as you could buy fuel. The heat source for this heat engine was gasoline or diesel fuel purchased from a growing network of filling stations. The oil industry grew up with the automobile industry, both eventually becoming the largest industries in the world. Of the top ten corporations in the 2005 Fortune Global 500 list, four were auto manufacturers and four were oil companies.[13]

A car represented freedom because it allowed individuals to direct unprecedented amounts of energy for their own purposes. A driver could control the equivalent of a hundred or more horses simply by pressing down with one foot. The personal mobility made possible by the car transformed human interaction.

The internal combustion engine reinvented agriculture as well. For example, the earliest combine grain harvesters were horse drawn, and handled cutting, threshing, and winnowing all at once. Such a machine greatly cut down on human labor, but it required a team of up to forty horses to pull it, and several human handlers to coordinate so many animals. A 40-horsepower diesel engine soon replaced those forty horses. Modern combines range up to 600 horsepower (450 kilowatts).

The near-complete automation of farming tasks freed most of the population from heavy farm labor. In 1850 in the United States, 64 percent of the labor force worked on farms,[14] but that went down to just 1.6 percent by the year 2010.[15] Releasing most people from farm labor freed them to pursue education, business, and other social advancements.

The ease of operation of an internal combustion engine and the convenience of liquid fuel over coal also led to the internal combustion engine replacing steam engines in trains and ships. Trains today are powered by diesel engines burning liquid fuel to generate electricity that drives electric motors in the wheels. Almost all ships run on diesel engines today.

The light weight of an internal combustion engine made powered flight possible too. Flight requires plenty of power, but the early steam engines could produce only about 10 watts per kilogram of engine weight, while an internal combustion engine can produce 1,000 watts per kilogram.[16] The first commercial flights used propellor-driven airplanes powered by piston engines.

The jet engine, another kind of internal combustion engine, can produce 10,000 watts per kilogram. It made fast intercontinental travel possible, then commonplace. The physical interconnectedness of human civilizations that began with wind-powered sailing ships was complete, thanks to energy.

CHAPTER 5

Fossil Fuels and Heat Engines Advanced the Human Race

With the spread of the internal combustion engine, the human relationship with energy took a fateful turn. The 1800s had seen the gradual growth in using coal for energy, first as a replacement for wood energy in steam locomotives, and then to power the new electricity service. The internal combustion engine's thirst for liquid fuels drove the expansion of a second fossil-fuel industry—oil. Once we started on the path to fossil fuels, we dove into those treasure troves of energy with little restraint.

Fossil fuels allowed humans to tap into an energy source that was seen at the time as unlimited. Coal was simply dug out of the ground, and oil gushers symbolized a bounty of natural energy. We quickly began to exploit great quantities of energy using the newly developed heat engines. This flood of energy changed all aspects of human life. To see how much has changed, take a look at a snapshot of life before fossil fuels, in 1820, the year my tall clock started ticking.

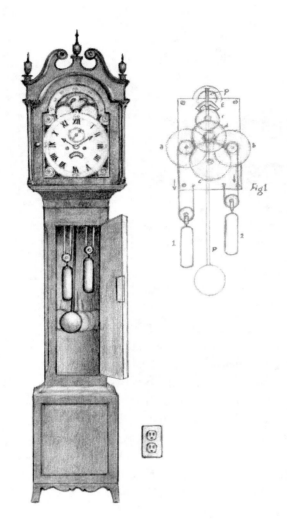

Figure 20. 1820 Clock

I inherited an antique grandfather clock from my father. Its impressive eight-foot-tall walnut case houses a brass clock mechanism powered by descending iron weights. It was built in 1820 by James Tanyard (I have the original receipt), and like most appliances of the time, the clock is hand powered. As long as I remember to wind the clock once a week to lift the weights inside, it keeps perfect time.

In the year my clock was built, US president Thomas Jefferson was still living, and he kept and wound a similar clock that is still preserved in his Monticello home. He heated his home with firewood, and cooked with firewood. When hot water was needed, he heated it on his wood stove. All that firewood required someone to cut, split, haul, store, and feed it into the fire. Water was drawn or pumped using muscle power.

Jefferson was wealthy enough to hire servants and field hands for most of this work. But some of that human muscle energy came from slaves, who in 1820 had not yet been freed. Even that champion of liberty participated in the slave-energy system, using hundreds of slaves to work his farms.

In 1820, electricity service was still sixty years in the future, so lighting came from oil lamps or candles All work in the kitchen and shop required hand tools and muscle power. Clothes washing was done by hand in a tub or stream, with a clothesline as the dryer. Without refrigerator or freezer, food that was not dried, salted, or fermented had to be eaten up quickly or it would spoil.[1]

Transportation on water was powered primarily by wind energy filling sails, though goods were often transported inland on canals with boats towed by horses or oxen harnessed to walk along the edge of the canal.

On land, horses provided most of the energy to transport humans about, either as individual riders or as passengers on carriages or coaches. Goods were transported on land in wagons pulled by horses, mules, or oxen. There were no railroads yet, as the first steam railroad service would not appear until five years later.

Compared to the physical requirements of life in 1820, our modern life is easy. We don't have to pump water by hand, labor in fields to raise food, or chop wood for heat.

Today, we dwell in homes that are kept warm in winter and cool in summer by automatic systems controlled by thermostats. We have hot and cold running water on demand. Light at night requires only the flick of a switch. Laborious tasks can be offloaded to electric motors or gas engines. Food can be secured simply by driving down to the local grocery store.

Modern life is characterized by ubiquitous electricity, fast transportation, and easy access to goods. All of these are possible because we learned to control the flow of energy beyond our bodily food energy.

Energy flows make our lives easier, safer, and more comfortable—if you are lucky enough to have access to energy flows. If not, then your life will not be so easy, safe, or comfortable. By any measure, energy raises the standard of living for those who have access to it.

Modern people are able to direct large flows of energy because of energy systems, developed during the 19th and 20th centuries, that could process fossil fuels. The most important of these was the heat engine. The heat engine in the form of steam engines enabled the automation of work in factories, which increased the availability of goods. The heat engine in the form of internal combustion engines made fast individual transportation possible, and automated most farm work. The heat engine in the form of steam turbines made electricity for power, light, and heat available to anyone connected to the grid. The heat engine in the form of jet engines made every corner of the world easily accessible to all who could afford a ticket.

Heat engines were so successful at directing energy for our purposes that they gradually took over more tasks and functions. Today most physical work is not performed by human laborers but by machines powered directly by heat engines or indirectly by electricity produced from heat engines. If you need a trench dug, you hire a backhoe operated by one person instead of a team of human shovelers. If you need to saw a board, you use a power saw instead of a handsaw. If you need a pencil sharpened, you use an electric sharpener instead of turning a crank. Manual labor is used only when a task is so complicated that a machine has not yet been invented for it (such as framing a house, and even then power tools provide assistance), when a machine is too expensive (in an African village, for example), or when work is done voluntarily (such as a personal garden).

Urban humans have largely abandoned the use of bodily energy to perform useful tasks. Almost all jobs that require heavy labor can be performed by a machine. Now we expend bodily energy in designated exercise sessions that simply dissipate the energy. Such exercise is necessary to keep our bodies healthy by imitating our former vigorous life of survival in the wild, when our only energy source was our muscles.

Remove heat engines, and we go back to life as it was in 1820. Perhaps some people would appreciate a simpler life, but few would be willing to do the work to sustain such a life. Children raised in modern society have no notion of what that life would be like. There is no question that modern humans take energy for granted.

The growth in the use of heat engines would not have been possible without fossil fuels. The abundance of fossil fuels and the relative ease with which they could be extracted and prepared for use made them cheap. Fossil-fuel energy was cheaper than human labor, cheaper than animal labor, cheaper than wood fuel shipped a long distance, and cheaper than whale oil. The low cost of fossil fuels drove their adoption for every energy purpose.

Today, fossil fuels and heat engines revolutionize life for those with access to them. The ground-shifting change is the ease with which individuals can direct larger flows of energy to meet their needs. Instead of directing a single horse for transportation, you can direct the equivalent of a hundred or more horses when driving a car. Pressing the up button on an elevator can lift a dozen people hundreds of feet in the air in seconds. Easy control of such large energy flows was unheard of in 1820.

Modern, high-energy civilization—marked by megacities, globalized economy, unprecedented levels of affluence, intensive transportation, instant communication, a surfeit of food, and the amassment of possessions—could not have arisen without the high energy densities of fossil fuels, portability of refined oil products, and superior flexibility of electricity.[2]

—Vaclav Smil, Distinguished Professor Emeritus,
University of Manitoba

Fossil fuels have become the foundation of society, but the civilization-threatening problems of global warming and ocean acidification are undermining that foundation. Since these problems emerge from the fossil fuels that prop up modern life, we must abandon either fossil fuels or our easy lives. Resolving this dilemma is the greatest test humanity faces today.

PART I: HOW WE GOT HERE

PART II
Our Energy Dilemma

Modern Energy Consumers

The individual human relationship with energy shifted dramatically when fossil fuels took over. As electricity and natural gas service spread from cities to suburbs to rural areas, and as automobiles and trucks took over transportation, the relationship that most people had with energy shifted from active participant to passive consumer.

The energy supply for a car comes in the form of gasoline. Your role is simple—watch the fuel gauge, and when it gets low, stop and fill up, paying for the gas with cash or card. Gasoline is simply a commodity that you buy as needed. Finding the cheapest price for gas is generally the extent to which you think about the process.

Paying a monthly utility bill is not much different. The bill shows how much electricity and, perhaps, natural gas you consumed, and you pay for that quantity. The major difference is that you pay after consuming the energy instead of before.

With modern energy systems based on fossil fuels, you no longer actively produce the energy you use. The primary reason is that you cannot. You cannot drill your own oil well and refine the crude oil into gasoline or dig your own coal and feed it to a steam turbine to generate your electricity.

Extracting and processing fossil fuels is not for amateurs. To produce oil, you must first locate an oil field, a task handled today by teams of petroleum geologists sounding for underground oil with specialized equipment. Then a drilling crew drills wells, with no guarantee of producing useful amounts of oil. If they do strike oil, it has to be transported to a refinery where a complex multistage process breaks it down into useful components.

In modern society, the complexity of energy systems is managed by professional energy corporations—oil companies, gas companies, coal companies, and utility companies. Their people manage the process of converting energy into a commodity to be metered out and sold as a consumer item.

Most energy consumers are satisfied with this arrangement, because the energy companies make it easy to use energy without you having to spend much effort on it, leaving you time for more important pursuits.

That convenience and simplicity, however, leaves modern citizens woefully ignorant of where their energy comes from. You do not know if the gasoline you buy comes from an oil well on the Texas plains, from a drilling platform in the North Sea, or from a region of conflict in the Middle East. Not only do you not know, you have no way of finding out. Gasoline does not come labeled with its country of origin.

If you are an eco-conscious commuter riding an electric bicycle, then you fare little better. When you pedal the bike, you might recognize that your muscle energy derives from the muffin you had for breakfast. But when you switch on the electric motor, the electrical energy, which is so clean compared to polluting automobile engines, is of unknown origin.

Your bike's electricity might come from a hydroelectric system, which uses the energy of falling water to spin a water turbine. If so, you could consider your bike to be gravity powered. That would be like always gliding downhill, something all bicyclists dream of.

If your electricity comes from a coal-fired power plant, you could consider your electric bicycle to be coal powered. Of course, you are not towing behind your bike a messy coal-fired power plant, but you are being pushed along by energy in the battery that was generated from coal. Effectively, that makes the bike coal powered.

Some utility companies operate nuclear reactors to generate electricity. If your electricity comes from a nuclear power plant, you could consider your electric bicycle to be nuclear powered. Imagine that.

You won't find a nuclear-powered bicycle in nature, but you can find one among the many energy devices available to modern energy consumers.

Modern energy directors

As a modern energy consumer, you might not know where your energy comes from, but you do play a crucial role in controlling where it goes. Every individual engages energy systems of many different kinds to carry out their daily activities. Traveling to work, cooking dinner, or watching television all require attention to an energy system that supports that activity.

In a way, it is amazing that human beings can control energy the way we do. After all, energy powers everything in our universe, from supernova star explosions down to living cell metabolism. We have learned ways to direct this fundamental force of nature to perform services for us on demand, and in sufficient quantities to do a job, but not so much that it gets out of control and destroys things. Energy—the Grand Master Mover and Shaker of the Known Universe—can be controlled by ordinary people. That's remarkable.

The energy systems that most of us use do not require that we understand their inner workings. Energy is powerful stuff, but it has been tamed by scientists and engineers to do our bidding in energy systems that are safe and easy to operate.

Most of us are not out there inventing new energy systems. Creating an energy system for the first time is not easy, yet once an energy system has been shown to work and produce some benefit, people have shown a remarkable ability to copy it and pass it along to others. The first one to control fire made a great breakthrough, but it was only a matter of time before everyone was using fire routinely.

What is energy?

So what exactly *is* energy? How much is known about the nature of energy itself? If you ask the experts in energy, the physicists and engineers, they will most often tell you that *energy is the capacity to do work*. In this context, *work* means applying a force to move some matter, so defining energy as the capacity to do work means energy is something that can be transformed into kinetic energy. But that just tells you what energy sometimes *does*, not what it essentially *is*. If you press the energy experts for more depth, they will finally admit that they don't really know what energy is. Nobel physicist Richard Feynman[1] said:

> It is important to realize that in physics today, we have no knowledge of what energy *is*.

So the people with the most knowledge of energy cannot tell us what it is. Energy is so fundamental that there is nothing more fundamental with which to explain it. That leaves energy as a mystery yet to be solved.

In modern times, ordinary people drive cars routinely and use electrical appliances that handle great amounts of energy. We become aware of the energy that's in play only when something goes wrong, as in a car out of control or an exposed electrical wire.

When you use an energy system, you are directing a flow of energy. For example, when you drive a gas-powered car, your foot presses down on the accelerator pedal to make the car go. Your action meters some gasoline into the combustion chambers of the engine, where the gasoline chemical energy is transformed into heat energy to push the car forward. Your simple action of pressing your foot down a fraction of an inch causes a vehicle weighing several thousand pounds to acquire a great deal of kinetic energy, setting the car in motion.

You did not personally push the car to get it moving; you directed a flow of energy to do so. You also did not personally measure out the gasoline and push it into the cylinders; that was handled by the car's automated fuel system designed by engineers. Your role was just pressing the accelerator pedal with your foot.

With that same foot, you apply the brakes to slow the car and bring it to a stop. Now your foot is engaging a different energy system, the brake energy system whose purpose is to transform the car's kinetic energy into heat energy. You don't need to skid your feet on the ground to slow down, nor do you need to know how the brakes work. Your role in this energy flow is just pressing the brake pedal with your foot. Overall, your actions control how the car behaves.

This control of a car's energy systems is emblematic of humans directing energy flows. With little effort, a person can control a large energy flow. Of course, simply releasing energy randomly does not serve our purposes. Setting fire to a can of gasoline will not get you to your destination. The energy must be released in the right context to have its intended effect, and it must be kept under control to avoid unintended consequences.

That control comes from the human operator. Every day you operate energy systems in a controlled manner to achieve your purposes. You direct flows of energy to do things that your muscles alone could not accomplish. The great thing about directing energy flows is that you don't have to do all the work yourself.

In a sense, you are like a movie director, who guides resources for making a film. The director controls actors, camera operators, prop managers, lighting technicians, sound technicians and all their associated equipment to create an engaging film. The director need not know how to operate the camera, the lights, or the sound board, but just directs their application. We each direct many forms of energy to accomplish our own tasks.

Energy servants

In the days before fossil fuels, those with money hired servants to perform work. If the fossil-fuel energy consumed in America were magically converted to food energy, there would be enough food to feed seventy-five servants for each American. So fossil fuels provide each American with the equivalent of a dedicated crew of servants working for them to make their life easier. A household of four would command a staff of 300.[2]

Directing energy is something every human on the planet has in common. Even the billion or so people living in small villages without electricity use wood energy for heat and cooking, and animal energy for power. Rich or poor, modern or not, we all direct energy flows of some kind in order to move about, secure food, make things, and ensure comfort. The familiar energy systems you use daily are part of your personal relationship with energy. Without those systems, your life would be difficult or even impossible.

To accomplish ever-larger purposes, modern society has grown ever-larger energy systems based on fossil fuels. Fossil fuels made possible fully loaded superhighways, hundred-story skyscrapers, and 10,000-acre farms with no farmhouse in sight. Modern life has grown up in tandem with the growth of fossil fuels.

CHAPTER 7

Fossil Fuels Put Us in a Bind

The success of fossil fuels grew into a dependency on them. As a prime example, consider our food system in the 21st century. On the farm, fossil fuels run the farm machinery that prepares the soil, spreads fertilizer, weeds, and harvests. Nitrogen fertilizer is made from natural gas, and pesticides are largely made from oil. Fossil-fueled trucks, trains, ships, and even airplanes transport the harvest to great factories for processing. Food-processing plants consume large quantities of fossil fuels directly and in the form of electricity for cooking, cleaning, and bottling, freezing, or otherwise packaging the food. Fossil fuels power the transportation of the packaged food, sometimes over long distances, to stores to be sold. The stores consume fossil-fuel-generated electricity for lights and refrigeration. The consumer transports the food home in a fossil-fueled car, and stores it in a refrigerator kept cool with fossil-fuel-generated electricity until it can be cooked with fossil-fuel natural gas or fossil-fuel-generated electricity.

The classic study of energy use in modern agriculture showed that the energy consumed to produce the food is eight times as much as the energy in the food.[1] A more recent study found that the diet of an individual American is supported by energy equivalent to 528 gallons of oil per year. That would amount to nineteen units of fossil-fuel energy for each unit of food energy consumed.[2] While food remains the primary energy source for people, it is no longer the primary energy source for human civilization.

Most of those food-handling steps could not be done without fossil fuels, either directly or in the form of fossil-fuel-generated electricity. While not all electricity comes from fossil fuels, the majority does, and the removal of fossil fuels as energy sources for electricity would quickly show how dependent we are upon them. Even if no electricity were used in the food system, farm machinery and food transportation are completely dependent on fossil fuels.

How dependent are we? Consider a science fiction scenario where a space virus arrives on Earth, and this virus happens to eat the carbon in fossil fuels. The virus spreads quickly and consumes most coal, oil, and gas, as well as derivatives like gasoline and diesel fuel. The food system I described would grind to a halt. Subsistence farmers might not be affected, but everyone else would be in trouble. It is estimated that cities hold about three days' worth of food on grocery shelves.

Dependent on the Internet

One can see a parallel process in the more recent integration of the Internet into businesses. There was a time when they ran without the Internet, using telephones, paper memos, and surface mail to communicate and transact business. Few operate that way now. The Internet suddenly made communication faster and easier. For example, typed and mailed memos were replaced by instantaneous email. Such improvements in efficiency could not be ignored. Thus the Internet was adopted step by step by each business to make it more efficient and more competitive. Over time, businesses became critically dependent on the Internet, and now many experience great difficulties when the Internet goes down.

No one planned for or conspired to create our dependency on fossil fuels. Cheap fossil fuels created their own economic force that gradually led to their adoption, one decision at a time. Once fossil fuels proved to be reliable, the old ways were abandoned and eventually forgotten.

For two centuries fossil fuels have served humanity well, making life easier, making more food and goods available, and in general improving our quality of life. This growing dependence on fossil fuels was not seen as a problem because new discoveries were made every year and the system grew to be highly reliable.

But fossil fuels have many downsides. Most of these problems were small when energy use was small, so they could be ignored. For example, the newly introduced automobile was seen as a clean alternative to a manure-producing team of horses, especially in crowded city streets. The emissions from the car's tailpipe seemed to just disappear into the air without any harmful consequences.

Now that fossil-fuel energy use has grown huge, the problems have grown huge as well. Automobile air emissions only became noticeable when cities filled up with cars and the pollution darkened the skies. Climate change reached public awareness after a decade of record high temperatures.

Ten Reasons Why Fossil Fuels Are Not Good For Us Anymore

The number and size of the problems with fossil fuels have grown to the point where we must decide if their problems outweigh their benefits. It is useful to enumerate these problems so we can document why maintaining the current system is becoming untenable.

1. Fossil fuels are warming the planet.
2. Fossil fuels acidify our oceans.
3. The conventional sources are nearly depleted.
4. Fracking is now compulsory to maintain the supply of oil and gas.
5. Oil is not cheap anymore.
6. Oil imports hamper national economies.
7. Oil drives political conflict.
8. Fossil fuels destabilize world food prices.
9. Air pollution makes breathing hazardous.
10. Water pollution makes drinking hazardous.

Each of these reasons is explained more fully in the following sections.

1. Fossil Fuels Are Warming the Planet

The carbon dioxide emitted by burning fossil fuels traps heat on the Earth's surface by absorbing outgoing infrared radiation that would otherwise escape into space, cooling our planet in the process. Since the industrial age began, the atmospheric carbon dioxide concentration has increased from 280 parts per million to over 400 parts per million throughout the world's atmosphere. That 43% change has raised the average temperature of the Earth's atmosphere by 1.4 °F.

While that temperature change might not seem like much, it represents an enormous amount of extra heat energy added to the atmosphere, energy that alters climate patterns and powers extreme weather events. Natural processes that remove carbon dioxide from the atmosphere take centuries, so it is accumulating relentlessly the more we burn fossil fuels.

Even now we are seeing effects of this accumulation:[3]

- Ten of the last twelve years are ranked as the warmest on record, since global records began to be tabulated in 1880.[4]

- Average ocean temperatures have increased to depths of at least 10,000 feet, causing seawater to expand and contribute to rising sea levels.

- Mountain glaciers and snow cover have universally declined in both hemispheres.

- Arctic sea ice has decreased so much in summer that ships can now pass north of Canada for the first time in history.

- Evidence from plants and animals indicates that spring is occurring earlier and fall is occurring later, causing mismatches in reproductive cycles between species that are dependent on each other.

The consensus among climate scientists, as expressed in the Fifth Assessment of the Intergovernmental Panel on Climate Change (IPCC), concludes:

> Warming of the climate system is unequivocal, and since the 1950s, many of the observed changes are unprecedented over decades to millennia. The atmosphere and ocean have warmed, the amounts of snow and ice have diminished, sea level has risen, and the concentrations of greenhouse gases have increased.
>
> —Intergovernmental Panel on Climate Change[5]

The consequences of continued warming are predicted to be:[6]

- Permanent displacement of tens to hundreds of millions of people due to rising sea levels.
- A massive reduction of water supplies in some parts of the world, especially those that rely on the steady run-off of meltwater from glaciers and snowpacks.
- The spread of malaria, cholera, and other diseases whose vectors or pathogens are temperature- and moisture-dependent.
- Increased devastation from extreme weather events such as floods, droughts, wildfires, typhoons, and hurricanes.
- Significant loss of biodiversity.

Many factors besides carbon dioxide contribute to climate change, including some that induce cooling. For example, forest land cleared for farming reflects more sunlight, as do aerosols emitted by combustion, so both have cooling effects on the planet. But when all the effects are quantified, carbon dioxide is found to contribute fully 73% of the *net* force driving climate change.[7]

So it is not possible to stop human-induced climate change without dramatically slowing human-emitted carbon dioxide. The urgency derives from its permanence. Because the natural mechanisms that remove excess carbon dioxide from the atmosphere are extremely slow, carbon dioxide emitted today will stay up for hundreds or thousands of years. We cannot wait for the effects to get worse before taking action.

2. Fossil Fuels Acidify Our Oceans

The carbon dioxide released by the burning of fossil fuels for the last 200 years has not all stayed in the atmosphere. About a third of it has been absorbed by the oceans. That process has actually helped slow the effects of climate change by removing carbon dioxide from the atmosphere, but at the expense of harming the oceans.

The carbonic acid made by carbon dioxide makes the oceans more acidic. So far a 30 percent increase in surface acidity has been measured, and a 150 percent increase is expected by the end of the century. The acidification has a host of biological effects, the major one preventing shelled animals from forming their shells.

Ocean acidification has been blamed for causing oyster farms in Oregon and Washington State to fail. The oysters hatch, and the juvenile oysters appear to be fine, but they never form their shells. Marine scientists attribute the failure to the measured change in acidity along the coast.[8]

Tiny shelled animals called diatoms, and tiny floating snails called pteropods, also cannot extract calcium from acidified water to form their shells. These creatures form the bottoms of food chains for many species, and when you mess with the bottom of a food chain you mess with all the species dependent on it.[9]

Acidification also inhibits coral reef formation. Of the six field studies performed to date, all six showed a reduction in coral calcification.[10] Since about 25 percent of all ocean species spend at least part of their life in coral reefs, any further reduction in coral growth could have cascading effects on marine organisms, food webs, biodiversity, and fisheries.

Our carbon dioxide emissions are slowly poisoning the oceans. Adding acid is adding a kind of poison that kills by dissolving shells. The equivalent for humans would be the air around us becoming corrosive enough to dissolve our skin.

The combined effects of our carbon emissions are creating the conditions for mass extinction. In the history of the Earth, all five previous mass extinctions showed three symptoms: Warming, ocean acidification, and hypoxia (reduced dissolved oxygen). All three are present now due to carbon dioxide.[11] To make it worse, the current rate of change of global ocean acidity is unprecedented, a factor of 30–100 times faster than changes in the recent geological past. [12]

Ocean acidification is a separate effect from climate change, and its scientific basis is not in dispute. Weather is variable and unpredictable, so finding an overall global warming effect is difficult because it gets lost in the variability. Measurements of ocean acidity are precise, consistent, and very slow to change. Current changes in ocean surface-water chemistry can be unequivocally linked to increases in atmospheric carbon dioxide.[13]

Ocean acidification is the smoking-gun evidence that convicts fossil fuel emissions of harming the planet. You don't need to believe in climate change to accept that fact.

Ocean acidification is essentially irreversible. There is no practical way to remove the carbon dioxide from the oceans once absorbed, and it will take tens of thousands of years for natural processes to restore the balance.[14] Fixing it through artificial methods such as adding chemicals is unproven and impractical for entire oceans. The only way to stop ocean acidification is to stop emitting so much carbon dioxide.

3.   The Conventional Oil Sources Are Not Meeting Demand

Our most important fossil fuel, the one we cannot live without, is oil. The conventional method for producing oil is to drill down into a reservoir and pump out the oil. When the well is sufficiently depleted that pumping costs exceed the value of the extracted oil, the well is abandoned. This method, which dates back to the late 1800s, met the world's oil needs until recently.

World oil consumption has risen steadily over the decades, with a few temporary declines usually caused by economic downturns. Recent growth in energy consumption has come from the emerging economies of India and China, which are rapidly industrializing and adopting Western patterns of energy use. Historically, rising demand spurred searches for new oil fields to meet that demand. That is still the case today, but the searches are less fruitful now.

The following figure illustrates the major problem with the world's oil supply. It shows the rate at which new oil produced by conventional methods was discovered from 1930 to 2008, with projections beyond 2008 based on current trends. It is clear that despite occasional upticks from major new finds, the trend of discoveries after the 1960s has been downward.

This decline is even more alarming when you consider that the technology for finding oil has improved enormously during that time. Satellite imaging, gravity surveys, magnetic surveys, and seismic soundings subjected to computer analysis are among the sophisticated tools availa-

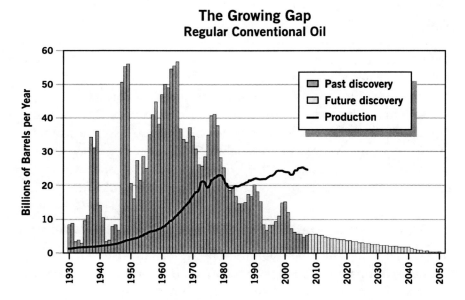

Figure 21. Declining conventional oil discoveries

The vertical bars show the worldwide rate of new conventional oil discoveries in each year, excluding fracking and oil sands. both historical and projected. The line shows the rate of oil consumption. New discoveries are not keeping up with consumption.

ble today. All continental areas have been surveyed except the Antarctic, which is protected from oil drilling by international treaty until 2041. Only offshore and Arctic areas hold much hope for big new discoveries of conventional oil fields.

Today, 80 percent of the world's oil comes from fields discovered before the early 1970s, and many of those fields have declining production.[15] For example, the Ghawar oil field is Saudi Arabia's largest, producing 70 percent of Saudi oil output. But that field started production in 1951, and is now declining at 8 percent per year, even though they inject millions of barrels of seawater to drive the oil out of the ground.[16]

4. Fracking Is Now Compulsory to Maintain the Supply

With conventional oil in decline, drillers have had to resort to new and unconventional methods to meet the demand. The most important of these is hydraulic fracturing, or *fracking*.

Fracking is employed when oil or natural gas is trapped in tight rock, such as shale. Drillers force a mixture of water, fine sand, and chemicals into a well under pressure high enough to crack the rock. The sand grains suspended in the fluid prop open the cracks, and the chemicals facilitate the flow of oil or gas.

In the United States, most easily pumped conventional oil and gas wells have been tapped out, [17] and until recently oil imports were making up the difference. But the country has large reserves of underground shale oil and gas that can be extracted using fracking, so the US is now experiencing a fracking boom. In 2013, about one-third of the oil and gas produced in the United States came from fracked wells, up from almost none in 2000. The levels are expected to reach 50 percent by 2040.[18]

The consequences of the US fracking boom include lowered energy prices from the new supplies, and reduced imports of foreign energy. But the growing dependence on fracked fuels will make it more difficult to address the emerging problems of the fracking process.

For example, the new gas supply enabled US electric utilities to switch from coal to natural gas to generate power. Because natural gas emits less carbon dioxide per unit energy than coal, this shift was seen as a positive move for controlling greenhouse gases. That gain is being offset by leakages of gas from the drilling and transport operations.[19]

Because methane is eighty-four times as potent a greenhouse gas as carbon dioxide over a twenty-year period,[20] even small leakages can negate the gains from switching fuels. One NOAA report found that methane leakage amounted to 6–12% of a Utah field's entire gas output,[21] but it is not yet clear if such leaks are due to poor operational practices that can be corrected.

Fracking operations have also been implicated in contamination of drinking water due to fracking chemicals leaking into groundwater, surface-water pollution from extracted fracking materials accidentally spilling from their impoundments, and enhanced earthquake activity triggered by the drilling and pressure operations.[22]

Despite these emerging problems, the US government continues to support fracking by exempting its operations from the Clean Water Act, the Clean Air Act, the Safe Water Drinking Act, and the National Environmental Policy Act.[23] These exemptions were put in place to encourage the growth of domestic fossil-fuel supplies. Since the country is now dependent on fracking, the exemptions leave American citizens with little recourse if anything goes wrong.

5. Oil Is Not Cheap Anymore

If one defines "cheap oil" as oil priced below $30 per barrel, then the 20th century was the era of cheap oil. That era ended when the 20th century ended.

In 1947, the price of crude oil on the world market was $23 per barrel, and in 1999 it was again $23 per barrel (both expressed in constant 2014 dollars).[24] During that period the price had rollercoastered up and down responding to world events, but, remarkably, the price at the end of the period was the same as at the beginning.

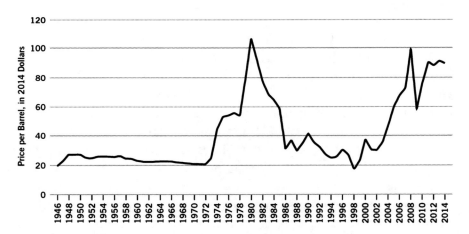

Figure 22. Crude oil price history (in constant 2014 dollars)[25]

Prior to 1973, the price of oil never rose above $30 per barrel (as measured in 2014 dollars). In 1973, the Arabic oil-exporting countries enacted an oil embargo against the United States that plunged the world oil markets into turmoil, doubling the price of oil. The Iranian Revolution of 1979 triggered another oil crisis, driving the price to over $100 per barrel. The economic recession of the early '80s and energy-efficiency measures triggered by the high prices resulted in reduced demand and a return to $30 per barrel in the 90s. Since 2000, oil prices have been rising.

Throughout the 20th century, oil was abundant and available from many countries, and competition served to keep the price down. If one country tried to sell oil at a higher price, another country with low production costs would produce more to undercut them to sell more of their oil. The price of oil trended downward to approach the cost of producing the oil, as that is the lower limit for oil companies to operate.

Oil in the 21st century presents a different picture. Starting in 2004, oil prices began climbing steeply as demand began to exceed supply, partly due to the growth of demand from the emerging economies of China and India. By 2008 the price had soared to $99 per barrel, as shown in the following figure. After a sharp drop in 2009 due to the worldwide recession, oil prices have bounced around between $60 and $100 per barrel.

A higher price for oil has a side effect of making more oil available for sale. When the selling price of oil is high, oil in the ground that is difficult and thus costly to extract becomes profitable. There are many oil fields that are currently considered "depleted" because their cost of producing a barrel of oil is higher than the price they get for that barrel. Such fields could become productive again when the price gets high enough.

Similarly, higher oil prices have enabled fracking operations to grow, even though fracking is more expensive than conventional oil production. Since the world is growing dependent on fracked oil to replace dwindling supplies of conventional oil, the price of oil must remain high to support the costs of fracking. When the price of crude oil in the U.S. declined to $50 per barrel in late 2014, many fracking operations shut down.

So the era of oil will be extended as the price of oil goes up, but those "new" sources are unlikely to drive the price back down to $30 per barrel.

6. Oil Imports Hamper National Economies

Higher oil prices create great economic stress for oil-importing countries, which must export money to pay for that oil. The exported money leaves their national economy instead of recirculating in it. This acts as an economic drain on such countries.

With higher oil prices, this problem worsens. Prior to the year 2000, the U.S. never exported more than $65 billion per year to pay for its oil imports. That shot up to $334 billion in 2008 when oil prices surged.[26]

For an economy struggling to get out of a recession, such huge oil trade deficits divert capital from new investments.

Worldwide, ninety countries are net energy importers, while forty-two are net energy exporters. Of the top ten energy-using countries, eight are net energy importers (Canada and the Russian Federation are net exporters).[27] Thirty-two countries import more than 80 percent of the energy they use.[28]

Such an economic drain threatens the survival of some nations.

> A number of these states have been identified by outside experts as at risk of state failure—the Central African Republic, DROC, Nepal, and Laos, for example. States characterized by high import dependence, low GDP per capita, high current account deficits, and heavy international indebtedness form a particularly perilous state profile. Such a profile includes most of East Africa and the Horn.
>
> —U.S. National Intelligence Council[29]

7. Oil Drives Political Conflict

The dependency of many national economies on imported oil inevitably leads to political conflict. Oil runs 95 percent of the world's transportation, and transportation of people and goods is essential to all functioning economies. The global economic system is wholly dependent on a constant supply of oil, but that supply is more at risk today than ever before.

With new discoveries not matching the surging demand, the oil supply chain is strung tightly, with little extra production capacity available to correct for supply disruptions from world events. Much of the world's oil comes from the Middle East and Northern Africa, two regions rife with conflict. The price of oil always spikes higher on bad news from those areas.

Even good news can raise oil anxieties. Oil-importing nations may have cheered the democracy movements of the Arab Spring of 2011, but for decades those same nations propped up undemocratic dictators in oil-exporting countries. The Shah of Iran, Saddam Hussein in Iraq, the King of Saudi Arabia, and others received billions of dollars' worth of armaments to protect the supply of oil, with the side effect of suppressing all dissent in those countries. Replacing a stable tyrant with an unpredictable democracy injects a level of uncertainty and risk into the world oil market.

A political dispute also led to the cutoff of natural gas to Europe in 2009. Europe imports much of its natural gas from Russia, delivered through pipelines that pass through Ukraine. In January 2009, a dispute between Russia and Ukraine prompted Russia to halt the flow of gas through the pipelines, leaving a dozen European countries shivering in the winter cold.

Oil transportation has similar chokepoints. Half the world's oil production is delivered in oceangoing tankers. Any of the key shipping routes such as the Strait of Hormuz, the Strait of Malacca, the Suez Canal, or the Panama Canal could be disrupted through war, accident, or terrorism. According to a publication of the US Department of Energy, "The blockage of a chokepoint, even temporarily, can lead to substantial increases in total energy costs."[30]

Oil is so essential now that national governments feel justified or even compelled to go to war over it.

> Let our position be absolutely clear: An attempt by any outside force to gain control of the Persian Gulf region will be regarded as an assault on the vital interests of the United States of America, and such an assault will be repelled by any means necessary, including military force.
>
> —President Jimmy Carter, State of the Union address,
> January 23, 1980

This policy firmly expressed itself in two recent wars over oil, which could be called Oil War I and Oil War II. Oil War I started when Iraq invaded Kuwait in 1990. Iraq accused Kuwait of stealing its oil by slant-drilling across its border. Iraq also claimed Kuwait was perpetrating economic warfare by keeping oil prices low through its high oil production, which made it difficult for Iraq to repay its debt from the Iran-Iraq War. What would have normally been a local conflict assumed global proportions because both countries produce oil for the world market. The United States led a coalition of forces to free Kuwait in 1991, restoring to power the Kuwaiti monarchy.

George W. Bush's invasion of Iraq in 2003 could be called Oil War II. The original excuse for the war proved false, as no weapons of mass destruction were ever found. In a Sky News interview in 2003, Sir Jonathan Porritt, head of the Sustainable Development Commission which advised English Prime Minister Tony Blair's government on ecological issues, said the prospect of winning access to Iraqi oil was "a very large factor" in the allies' decision to attack Iraq in March. "I don't think the war would have happened if Iraq didn't have the second-largest oil reserves in the world," Porritt said.

When nations compete for oil to keep their economies stable, political conflict becomes inevitable. In 2005, the Chinese National Offshore Oil Corporation attempted an $18.5 billion takeover of US oil company UNOCAL. The United States government rejected the deal because it threatened US national security.

Even without war, oil distorts politics. When the United Nations imposed an arms embargo on Sudan for its human rights violations in the Darfur region, China undermined it by continuing to supply weapons.

8. Fossil Fuels Destabilize World Food Prices

The world's food supply is now tightly coupled to fossil fuels. As the world's population grew in the 20th century, the so-called *Green Revolution* enacted in developing countries managed to increase food production sufficiently to stay ahead of the growing demand for food. The Green Revolution was based on high-yield seed varieties that required irrigation, nitrogen-based fertilizers, and pesticides, all of which consume fossil-fuel energy. Most countries have now shifted from independent subsistence farming, which could not support the growing populations, to farming methods dependent on fossil fuels.

That dependency links the price of food to the price of fossil fuels. When oil prices rose sharply just before the 2008 recession, food prices jumped as well, sparking riots in numerous countries.[31] When the price of oil fell during the recession and then climbed back, food prices followed closely.

Current trends will further cement the link between food and fossil fuels.

- Agricultural research has declined worldwide,[33] so few alternatives to energy-inefficient farming practices are being developed.

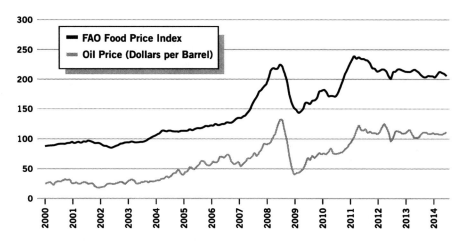

Figure 23. Food prices track oil prices[32]

Because modern farming depends heavily on fossil fuels, the cost of food reflects the cost of fossil fuels, shown here using oil as the price gauge.

- World population continues to grow, so nations are pressured to expand energy-intensive Green Revolution farming practices to replace subsistence farming.
- The continuing rural-to-urban migration in developing countries removes farmers from the land, forcing more fossil-fueled mechanization to feed the growing urban populations.

Food systems that are based on nonrenewable resources like fossil fuels are by definition unsustainable in the long run.[34] In the short run, the wild price fluctuations common in the world oil markets will guarantee widespread food insecurity.

9. Air Pollution Makes Breathing Hazardous

The exhaust from burning fossil fuels is a hazardous mix of pollutants. Although these pollutants become diluted when mixed in the atmosphere, continuous output from thousands of power plants and millions of vehicles overwhelm the ability of the atmosphere to disperse them. The list of air pollutants routinely emitted by burning fossil fuels includes:

- Sulfur oxides, which form when the sulfur impurities found in coal and oil are burned. Sulfur oxides are health hazards, and when mixed with water vapor in the atmosphere, they form acid rain that acidifies lakes and streams.
- Nitrogen oxides are produced from the nitrogen found in combustion air reacting with oxygen at high temperatures. Nitrogen oxides form the brown haze over polluted cities, and react with sunlight to produce ground-level ozone. Both nitrogen oxide and ozone are hazardous to lungs.
- Carbon monoxide results from incomplete combustion of the carbon in fossil fuels, and is highly poisonous. Most carbon monoxide comes from cars and trucks.
- Particulates are unburned residues from fossil fuels and traditional biofuels that are tiny enough to float in the air. When the finest particles are inhaled, they can lodge deep in the lungs and create respiratory problems.
- Toxic organic compounds are emitted by burning gasoline or diesel fuel. These compounds include benzene and formaldehyde, both proven to cause cancer.[35]
- Mercury, found in some coals, is released into the atmosphere when the coal is burned. When it settles out of the atmosphere it is taken up by fish and shellfish, where it accumulates as methyl mercury, which is highly toxic to developing fetuses and children. About 40 percent of US mercury emissions come from coal-burning power plants.[36]
- Radioactive uranium and thorium, found in most coals, exposes populations downwind from coal-fired power plants to more extra radiation than a nuclear-powered plant.[37]

Each item above has been demonstrated to be harmful to people, especially the elderly and young. Taken all together the harm is magnified. The World Health Organization estimates that outdoor air pollution caused 3.7 million premature deaths worldwide in 2012.[38]

10. Water Pollution Makes Drinking Hazardous

Oil contains dozens of toxic components, including over a hundred polycyclic aromatic hydrocarbons (PAH), which can cause infertility and may stunt the growth of fetuses by damaging their DNA.[39] These chemicals are spilled into our waterways by the vast oil-production and delivery system.

- Oil and gas exploration and production routinely leak 11 million gallons of oil into the world's oceans every year.[40] Such leaks are occasionally boosted by spectacular failures like the blown-out oil well of the BP Deepwater Horizon rig in 2010, which released over 200 million gallons of crude into the Gulf of Mexico.

- The world-spanning oil transport system routinely leaks 44 million gallons of oil each year, mostly from tanker ships.[41] As with drilling, such routine leakage is occasionally supplemented by tanker accidents like the *Exxon Valdez* spill in Alaska or the *Amoco Cadiz* off the coast of France.

- Consumers of oil products are the surprising leaders in oily water pollution. A car oil change that gets dumped instead of recycled, an engine crankcase that drips oil, or a two-stroke engine that emits oil droplets in its exhaust are multiplied by the millions worldwide. Most of these small leaks occur on land, but rain washes the oil into streams, then rivers, and eventually the world's oceans, polluting all of them in the process. These diffuse sources are estimated to add up to an enormous 140 million gallons of oil being absorbed by the oceans every year.[42]

Coal results in water pollution when rain leaches through the exposed waste piles that coal production leaves behind, including:

- Strip-mined soil and rock residue, which is usually acidic.

- Mountaintop-removal residues leaching into streams and groundwater. In this mining method, an entire mountaintop is blasted apart, the coal separated out, and the rest dumped into a mountain valley, filling it with polluting waste. As of 2011, Appalachian forest areas larger than the states of Delaware and Rhode Island combined have been converted to surface mines.[43]

- Coal slurry, a mix of solid and liquid waste from cleaning coal, includes a long list of harmful chemicals from the coal and the cleaning fluids.

- Ash slurry, the leftovers from burning coal, is typically laced with toxic heavy metals like mercury, selenium, arsenic, chromium, and cadmium. A dike failure in Tennessee in 2008 dumped over a billion gallons of toxic ash slurry into the Emory River and surrounding lands.

Chemicals used to clean coal have also polluted waterways. In January 2014, an estimated 10,000 gallons of the coal-cleaning chemical MCHM spilled into the Elk River in West Virginia, rendering undrinkable the water supply for 300,000 people.[44]

Even extracting relatively clean natural gas contributes to water pollution. Natural gas obtained by fracking can pollute groundwater. A 2011 congressional investigation reported that oil and gas companies injected over 32 million gallons of diesel fuel or fracking fluids that contained diesel fuel in 20 states between 2005 and 2009.[45] Diesel fuel contains toxic chemicals including benzene, toluene, ethylbenzene, and xylene, all of which pose public health risks.[46]

The EPA is concerned that these chemicals will migrate to nearby sources of drinking water. An EPA investigation of the Pavillion gas field in Wyoming found, "Overall, 17 of 19 drinking water wells sampled in January 2010 show detections of total petroleum hydrocarbons,"[47] though the EPA did not reach any conclusions on how the chemicals occurred in the wells.

Chronic exposure to polluted drinking water can affect the body in many ways. For example, coal mining through mountaintop removal has been associated with a higher incidence of birth defects in the region where the mining takes place. Rates were significantly higher for six of seven types of birth defects studied.[48]

A Poor Fit

Taken together, these ten reasons indicate that there are problems everywhere with fossil fuels. Almost all aspects of fossil fuels are potentially dangerous, damaging to the environment, politically charged, or economically risky.

When humans started using fossil fuels, they were a good fit for our energy needs. The use of coal stopped the devastation of forests for fire-wood, the introduction of kerosene saved the whales, and gasoline-powered cars kept horse manure off the streets. When fossil fuels were still new and their use limited, their problems were either small, hidden, or not yet known. They would have stayed that way except the use of fossil fuels kept growing, so the problems grew with them.

The future will see the problems grow even worse if current trends continue. If you examine the distribution of fossil-fuel use around the world, you'll find that rich nations burn the most to sustain the modern lives of their citizens. Yet rich nations have reached somewhat of a saturation point, where all energy needs are met and growth of fossil-fuel use has slowed considerably. The growth in its use today is mostly in developing nations such as China and India as they try to raise their standard of living to match the rich nations.

Consider what would happen if everyone on the planet raised their fossil-fuel energy use to match that of Americans today. Carbon dioxide emissions would quadruple,[49] carbon dioxide concentrations would reach 800 to 1,000 parts per million, and temperatures would rise 8 °F by the end of the century.[50] Such levels of carbon dioxide in the atmosphere would so drastically alter climate and oceans that the survival of human civilization would be at risk.

Staying on fossil fuels is a lose-lose situation. If we stay on fossil fuels and grow them to meet the world's growing energy demand, we will further harm ourselves with climate change and ocean acidification. If we stay on fossil fuels and do not grow them to meet the increasing energy demand, then we hobble our economies and create conflicts between rich and poor nations. As supplies are drawn down and shortages begin to appear, we can expect increasingly desperate competition and chaos.

Staying with fossil fuels is no longer an option if we want a stable world economy and improvements in the standard of living for everyone. Fossil fuel use will always be inhibited by our need to reduce carbon dioxide output. Continuing with fossil fuels will be like driving a car while pressing both the brake and accelerator pedals at the same time. We won't make much progress, and it will hurt a lot.

PART II: OUR ENERGY DILEMMA

CHAPTER 8

Crucial Decision Point on Energy

We have reached a crucial decision point on energy. We no longer face the question of *whether* we will stop using fossil fuels, but *when* and *how*.

One might ask: Why now? We have been using fossil fuels for over 200 years, so why are they suddenly so much of a problem that we must stop using them? The answer is scale and accumulation.

The fossil-fuel industries have grown almost continuously for 200 years. The scale of our energy use has increased so much that we now load 100 million tons of carbon dioxide into our atmosphere *every day*. Our atmosphere and oceans simply cannot take it. They cannot process that much new carbon dioxide, and so it accumulates over time. Enough time has now passed that the accumulation has altered the composition of our atmosphere and the chemistry of our oceans. Only now are the effects of those changes becoming obvious.

In a sense, we have lost control of our grand energy machine. The human species gained advantage over other species by controlling energy, that is, by applying energy flows strong enough to meet our needs but not so strong as to create havoc. Science and engineering have given us complete control over the application of energy, but we are still learning to control the *consequences* of our energy use.

The major consequences that we do not yet control are global warming and ocean acidification, both resulting from carbon dioxide emitted by burning fossil fuels. These may have originally been excused as *unintended* consequences, but now they are *known* consequences. If we willfully continue with fossil fuels despite knowing the results, we effectively make them *intended* consequences.

No nation would dare put forth a policy calling for more global warming and ocean acidification. All nations agree that these must be arrested if we are to have a secure future, yet the paltry actions taken to date have allowed worldwide carbon dioxide emissions to *accelerate*, not diminish.[1]

There are those who call our relationship with fossil fuels an *addiction* because we cannot seem to quit something we know is harming us. Addiction is far too mild a word for our situation. Addicts can quit their addiction and still live, but if modern society simply quit fossil fuels, it would not survive.

Abandoning fossil fuels today would crash our civilization. Fossil fuels power agriculture, industry, transportation, and commercial operations, either directly or through fossil-fuel-generated electricity. Our transportation system alone is 95 percent dependent on oil, and transportation is necessary for all trade and commerce. Not having oil to power our agriculture systems would lead to food shortages, mass starvation, and a catastrophic breakdown of society.

So we are more than addicted to fossil fuels, we are *wholly dependent* on them. In our current state, we literally cannot live without fossil fuels.

Modern society faces a terrible dilemma today. We are utterly dependent on fossil fuels, yet we know we must stop using them. This dilemma is the source of our paralysis on controlling climate change and ocean acidification.

If we acknowledge that we are dependent and not just addicted, then efforts to simply reduce fossil-fuel use without substituting something else will prove to be futile. You may as well ask a person to go without food.

The task before us is to decide how to *replace* fossil fuels. We are fortunate that at this point in time when we need to replace them, we actually have several possible choices.

Energy Choices for the Future

As Earth's energy-using animal, we need energy to survive, to run our economies, and to be safe and comfortable. The strength of our species comes from our relationship with energy—that is, our ability to direct arbitrarily large external energy sources. But that energy does not necessarily have to come from fossil fuels. The solution to our dilemma is to replace fossil fuels with other sources of energy.

Since fossil-fuel energy is the foundation of our modern civilization, talk of changing energy sources makes some people nervous. They would agree with the statement "We need fossil-fuel energy to keep our economy running." That statement is half true. It is true that to keep our economy running we need energy. That is indisputable because nothing happens unless energy flows, but it is false to say that we must burn *fossil fuels* to provide that energy. Other energy sources can work as well, even better.

Fortunately, we do have choices. Today's generation of scientists understands energy like no previous generation, and has developed many new kinds of energy systems. These need to be weighed against each other to see which best solves all the problems presented by fossil fuels, and avoids new problems.

Changing energy systems need not be traumatic, as demonstrated by previous transitions. We changed from human muscle power to animal muscle power for agriculture and land transportation. We switched from human rowing to wind power for water transportation. We transitioned from whale oil to kerosene to electricity for lighting. And we transitioned from wind and water power to heat engines fueled by fossil fuels to generate electricity and to power transportation. We need not fear energy transitions, because we have experienced them in the past, and mostly benefitted from them.

History has shown that changing energy systems is possible, and even desirable. Once you know energy transitions are possible, the concept of leaving fossil fuels behind creates less anxiety and more hope for the future.

This is the first time that changing energy systems will be an intentional choice made by human civilization. Previous transitions occurred because of a single discovery. Any new energy invention was put to use if it made life easier, without much consideration of negative consequences. The early adopters of coal-fired steam engines were not presented with a choice of new energy systems, just a new invention that could replace the old horse-powered system.

Today, modern technology and scientific understanding of all energy forms have made available to us many new energy inventions, from which we get to choose. This time we have better knowledge of the potential consequences of our choices.

The decisions we make today will have long-lasting effects. We must make our energy decisions carefully, using the widest zoom mode on our energy scope so we can see how new energy sources are connected and operate in our world. In addition to seeing what good a new energy system can do, it is important to evaluate what harm it might do.

Here are the two critical questions we must ask of any new energy system before we embark on a long-term relationship with it:

- Is it free of carbon dioxide emissions?
- Is it big enough to replace fossil fuels?

Any new energy source must be able to answer yes to both questions if it is to significantly replace fossil fuels. For example, replacing oil with synthetic oil made from coal might be big enough, but it would not be free of carbon dioxide emissions. Likewise, geothermal energy, in which electricity is generated by steam from a geological hot spot as is done in Iceland, is free of carbon emissions, but is not big enough to replace fossil fuels.

If an energy source gets past this initial screening, it must also satisfy a few other requirements:

- Environmentally friendly, so we avoid trading fossil fuel problems for new problems.
- Politically acceptable, to preserve social stability in our world.
- Economically possible, so our economies can thrive and not be smothered.
- Long lasting, so we don't have to do this again in a few years.

It is doubtful that any energy source can fully meet all of these requirements. The best we can do is compare the various energy sources to all of these criteria and try to choose the best match.

Today, there are basically four energy solutions with potential to meet these criteria:

Choice 1:	Stop using modern energy.
Choice 2:	Continue to use fossil fuels but capture and store the carbon dioxide output.
Choice 3:	Use nuclear energy.
Choice 4:	Use solar energy in all its forms.

Each of these potential solutions is discussed in detail in the following chapters.

PART II: OUR ENERGY DILEMMA

CHAPTER 9

Eliminate Modern Energy

We could eliminate fossil fuels by slowly abandoning their use and going back to the days before we used them. This is not really a new energy system, but a reversion back to a way of living that did not require fossil fuels.

In 1820, almost no fossil fuels were in use. People lived, worked, ate, and enjoyed themselves without needing any. As described earlier in chapter 5, *Fossil Fuels and Heat Engines Advanced the Human Race* (page 65), firewood energy heated homes and water, horses transported people and goods on land, wind-powered sailing ships transported people and goods on water, and all goods were made using water power, animal power, or human power.

Of course, life was very different from today. There were no cars, trucks, trains, or airplanes, so most people did not travel very far. Agriculture required a lot of human power, so most people lived on farms. No one had electricity, so there were no electric lights, electric kitchen appliances, electric motors, or electronics of any kind. If you needed mechanical energy for tasks like pumping water and an animal could not provide it, you provided it yourself with your own muscle power.

Today, if you want to go somewhere, you get in your car, drive to your destination, park the car, and drive home when you are done. In the 1820 energy system, you instead go to the barn, saddle the horse, ride the horse to your destination, arrange care for the horse if you stay long, then ride home, unsaddle the horse, brush it down, and feed and water it. To care for the horse, you must also shovel manure, buy or grow the horse's food, transport the feed, transport and spread the bedding material, and call the veterinarian when needed. Horse transportation requires much more time and effort on the part of the energy user than a car does. The human relationship with energy was much more personal in 1820.

This is the kind of future that some people imagine when discussing climate change and the need to cut back on fossil fuels. To them, life without fossil fuels means returning to life as it was in the Wood Energy Epoch. But that vision only applies to people whose energy scope has a narrow focus and can see no alternatives.

Would the 1820 energy system meet our main criteria?

- Is it free of carbon dioxide emissions? Yes, because fossil fuels would not be burned, so carbon emissions would not be a problem.

- Is it big enough to replace fossil fuels? Not with our current population and rates of energy consumption. Consider just one case. Replacing US farm tractors in use today with horses would require at least 250 million horses, ten times the peak horse population in 1918. To feed the animals would take twice the total area of US farmland, an impossible demand.[1] With the 1820 solution, personal and industrial energy use would have to be scaled back to match the limited energy use of 1820.

Few people accustomed to modern conveniences would choose this solution to our fossil-fuel dilemma. It would only reach a global scale by a collapse of modern civilization. It is certainly not something we would choose for the advancement of human civilization.

CHAPTER 10

Clean Coal

Given our total dependence on fossil fuels, some people question the feasibility of transitioning to other energy sources. They ask the legitimate question: "If we can solve the existing problems of fossil fuels, could we avoid a major shift in our energy systems?"

Advocates of "clean coal" propose such a solution for the problem of carbon dioxide. The term *clean coal* refers to a coal-burning process that filters out and stores the carbon rather than emitting it into the atmosphere. Such a carbon capture and storage system (CCS) would eliminate coal as a contributor to climate change and ocean acidification. If we can get the energy without the carbon dioxide, the argument goes, we could continue to use fossil fuels with minimum disruption.

Three approaches for capturing carbon dioxide are being tried:

- Post-combustion, in which the hot exhaust gas is passed through a chemical process to separate out the carbon dioxide for storage.

- Pre-combustion, in which the coal is preprocessed with steam to gasify the coal into carbon dioxide and hydrogen. The hydrogen is burned for energy and the carbon dioxide is diverted to storage.

- Oxy-combustion, in which the coal is burned with pure oxygen instead of air. That eliminates the nitrogen found in air from the exhaust gas, leaving almost pure carbon dioxide to be cooled and stored.

All of these processes have been demonstrated on a small scale, but never in a full-scale coal power plant. The US government is attempting to kickstart the technology by funding the FutureGen 2 project, which would convert an existing 200 MW coal-fired power plant in Meredosia, Illinois, to oxy-combustion. When completed in 2015, its carbon dioxide exhaust gas will be compressed, transported by pipeline to a storage field, and injected underground.

Because the captured carbon dioxide must be transported by pipeline to a storage field, this technology can only be applied to large stationary sources of carbon dioxide, such as power plants. It could not be applied to moving sources such as cars and trucks, nor to the myriad small heating systems that burn fossil fuels. So only 20–40 percent of global fossil-fuel carbon emissions could be suitable for capture.[1] That figure could be raised by converting transportation to use electric motors, and using electricity as a substitute for fossil-fuel heating.

The collection process can capture 85–90 percent of the carbon dioxide from a coal-fired power plant. But the overall system of capture, compression, transport, and injection requires 25 percent more coal energy to be burned than a plant without CCS, offsetting some of the gain. The system *increases* the carbon emissions from coal mining, processing, and transport because 25 percent more coal must be handled.[2]

Since none of the CCS systems have been built at a full scale, the costs of these systems cannot be stated with a high degree of confidence at this time.[3] Costs will be lower for new power plants designed from the beginning with CCS, compared to retrofitting existing plants. It is certain that the cost of electricity produced from a CCS plant will be higher, by 25–100 percent based on current estimates.[4]

The fate of the captured carbon dioxide remains the greatest unknown. The oil and gas industry has some experience with injecting it underground. Natural gas coming out of the ground is usually mixed with carbon dioxide, which is separated and reinjected into the wells, partly to dispose of it but also to pressurize the wells to drive more gas out.

Such operations have never had a need to monitor the injected carbon dioxide to see if it stays underground, or if it gradually leaks out over time. The fear is that carbon storage will turn out to be a carbon dioxide time bomb inflicted on some future generation. Just when they think they have climate change under control, carbon dioxide starts leaking out of the ground from thousands of storage sites, defeating the whole purpose of CCS. There is not enough scientific evidence to say with confidence that such leakage will not occur.

The long time frames introduce more uncertainty. For CCS to be effective at halting climate change and ocean acidification, the captured carbon must be isolated for at least a few centuries. As Lloyd's of London points out, over such long time scales companies can change ownership or go out of business. Unprecedented legal frameworks must be put in place to ensure continuity of the maintenance, so that injection fields are not abandoned.[5]

And the volumes of gas are unbelievable. Even if our carbon dioxide gas emissions were compressed to the maximum degree before injection, the volume would be almost fifteen times that of the world's oil production. So while one huge industry pumps oil from the ground, a much larger industry would be needed to inject fifteen times that volume of material back into the ground.[6] It is also not known if enough geologic formations suitable for carbon dioxide storage are available, and whether injecting that much high-pressure gas underground would trigger earthquakes or shift groundwater.[7] Some have proposed injecting it deep in the ocean, but that would accelerate ocean acidification.

Overall, how does CCS answer our primary questions about future energy sources?

- Is it free of carbon dioxide emissions? No, it reduces some stationary sources by 80–90 percent, but does not apply to small or moving sources. And there is the risk that the stored gas will leak from the ground in the future.
- Is it big enough to replace fossil fuels? There are sufficient reserves of coal in the world to meet current energy needs for many decades, but it can replace *all* fossil fuels only if all other energy applications are converted to electricity so they can be covered by CCS.

It would appear that carbon capture and storage is not up to the task of stopping carbon dioxide emissions. CCS systems have limited impact on carbon emissions, and risk having no impact at all if the carbon dioxide leaks out in the future. And while scientists urge us to start reducing carbon emissions now if we are to avert disasters, CCS systems will not be ready for full-scale deployment for many years.

> CCS as a magical technology that solves the carbon problem for coal plants is oversold. ... I think there is a lot to learn, and it is going to take us a lot longer for us to figure it out than a lot of us think.
>
> —Jim Rogers, former CEO of Duke Energy, which operates sixteen coal-fired power plants.[8]

CHAPTER 11

Nuclear Energy

Nuclear power was once viewed as the natural successor to fossil fuels. Even in the 1950s, energy professionals were aware that fossil fuels existed in finite quantities and would eventually need to be replaced. Nuclear proponents envisioned a future of clean, abundant nuclear power. That we could convert the horror of nuclear weapons to peaceful nuclear power plants was seen as an added bonus.

The nuclear power industry grew steadily in the 1960s and 70s and was on track to fulfill that vision. In 1979, that growth halted in the United States with the accident at the Three Mile Island nuclear power station in Pennsylvania. The partial meltdown and release of radioactive gases demonstrated that nuclear power was not as safe as promoted, and this downgraded public opinion toward nuclear power. Utility companies, faced with rising construction and operating costs of nuclear power, essentially gave up. During the thirty years following Three Mile Island, no new reactor orders were placed in the U.S., and 120 existing reactor orders were ultimately canceled.[1] Other countries continued to develop nuclear energy, but at a slower pace.

Nuclear power had been set to make a comeback in recent years. Concern over climate change renewed interest in nuclear energy to replace the burning of fossil fuels. A nuclear-powered generating station

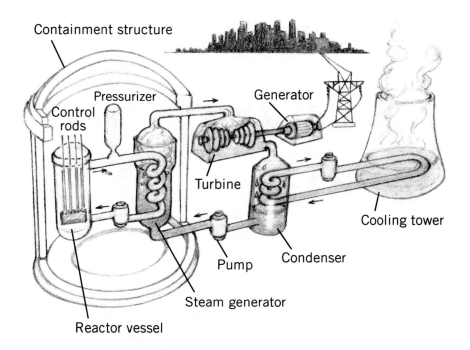

Figure 24. Nuclear Power Plant

A nuclear power plant has many similarities to a fossil-fuel power plant. A heat source boils water to make steam, which is fed into a steam turbine to turn an electrical generator. But instead of burning fossil fuels to generate heat, a nuclear power plant manages nuclear reactions to make steam.

emits no carbon dioxide during its normal operation. National governments seeking to meet carbon reduction goals saw nuclear power as a major option. Even environmentalist Stewart Brand, the editor of the Whole Earth Catalog, argued that nuclear power might be "the lesser of two evils" compared to climate change.[2]

We use nuclear power essentially as a fancy way to boil water, yet it differs from the fire our ancestors discovered in fundamental ways.

Nuclear energy itself derives from an entirely new force discovered only in the 20th century. Called simply the *strong force* by physicists, it holds every atomic nucleus together. The discovery of the strong force solved a conundrum for physicists, and led to the development of the nuclear bomb.

The conundrum comes from the way an atom is constructed. Recall a diagram of an atom, with electrons orbiting around a central nucleus, which consists of protons (with a positive electric charge) and neutrons (no charge). Since like electrical charges repel each other, scientists could not understand how all those positive protons could stay packed tightly together in a nucleus when the electrical forces should make it fly apart.

Their research led to the discovery of a new force that exists between protons and protons, and between protons and neutrons. The strong force attracts those particles to each other, but operates only at the extremely short distances within a nucleus. That strong force overcomes the electrical repulsion to hold the nucleus together.

That strong force also puts the nucleus under a great deal of tension, like a loaded spring. That tension is never expressed under most conditions, so the vast majority of atoms are stable. But it is possible to make that tension unstable, and that discovery led to the development of nuclear bombs, and later, nuclear power.

The trick is to use a neutron to make a certain kind of nucleus unstable. A neutron has no electrical charge, so if you shoot one at a nucleus, it is not repelled by the positively charged protons. If the neutron gets close enough, the strong force will suck it into the nucleus. In most nuclei, an extra neutron resides happily in its new home and does not destabilize the nucleus.

But in certain very heavy nuclei, an extra neutron can make the configuration unstable. For example, uranium is one of the heaviest elements because it packs so many protons and neutrons into its nucleus. All uranium nuclei contain exactly 92 protons, the number of protons that defines it as uranium. But the number of neutrons in each uranium nucleus can vary, from 141 to 146. Each variation is called an *isotope*, and you usually see an isotope number associated with it. The isotope number is the total number of protons and neutrons in that nucleus. So a uranium-235 nucleus has 92 protons and 143 neutrons (92 + 143 = 235), while uranium-238 has 92 protons and 146 neutrons.

If you drop a neutron into a uranium-235 nucleus, you form uranium-236, which is unstable. It will spontaneously break apart, typically into two large fragments and some extra particles. This split, or *fission*, of uranium-235 is the principle behind nuclear reactors in operation today. Each split atom results in several products:

- Two or more fragments of nuclei. These form new atoms when the fragments attract electrons.
- One or more free neutrons, which can then go on to trigger fission in other uranium-235 nuclei, creating the chain reaction in a nuclear reactor.
- Gamma rays, which are extremely high-frequency electromagnetic radiation that is immediately absorbed in the reactor core to make heat.
- Kinetic energy of the fragments, also contributing to the heat of the reactor core.

Note carefully that the list of fission products does *not* include carbon dioxide. There is no chemical burning of carbon fuels involved, so no carbon dioxide is formed as a byproduct. Getting energy without emitting carbon dioxide is the key to solving our energy problem.

And fission has a lot of energy compared to fossil fuels. If you burn a pound of coal, you release 11 million joules of energy. If you could split all the atoms in a pound of uranium-235, you get 32 trillion joules of energy, about *3 million times* the energy from coal.[3]

So with lots of energy and no carbon dioxide, nuclear energy seemed poised to resolve our energy dilemma. Some of your electricity may already come from one of the 400 existing nuclear power plants operating in the world today. Worldwide, about 14 percent of electricity comes from nuclear. In the United States it is 18 percent and in Europe it is about 25 percent.[4]

For nuclear power to eliminate fossil fuels, most energy applications would need to convert to electricity, because that is the only form of energy a nuclear power plant can provide.[5] Electric cars and trucks could run on nuclear electricity, as could all heating systems. Assuming all such conversions were carried out, how many nuclear power plants would be needed?

One estimate for total conversion from fossil fuels put the number at 10,000 reactors of the typical 1 GW size by 2060.[6] The construction rate would have to average 200 new plants per year for fifty years.

A more realistic estimate was developed by the International Energy Agency. Their scenario to halve global carbon emissions would quadruple the number of nuclear plants by 2050, requiring thirty-two large reactors to be built each year until 2050.[7]

These growth estimates run into a major barrier: There is not enough natural uranium in the world to power them all.[8] That many reactors would require breeder reactors, which use a different fuel cycle in which the waste products include another kind of nuclear isotope suitable for fission. The waste is processed to filter out the usable isotope to fuel other reactors. A breeder reactor can actually produce more nuclear fuel than it consumes, although of different kinds.

Using breeder reactors to generate more nuclear fuel has proven to be extremely difficult in practice. The United States shut down its only breeder reactor program in 1982 because of technical difficulties, spiraling costs, and fear of nuclear theft, since some of the elements produced by a breeder program are suitable for making nuclear bombs. In this terrorist-touched world, that's a frightening prospect.

Restarting a US breeder program would mean developing a new design, building and testing a prototype, developing a new commercial breeder reactor based on the prototype, building dozens of the reactors, building the reprocessing facilities to convert the output of the breeder into useful fuel, and building the hundreds of other reactors that can run on breeder fuel. It would take at least a decade before significant amounts of electricity could be produced from such a system.

Russia, India, and China are actively developing breeder reactors. Russia has had a 600-megawatt breeder reactor operating since 1980, and is developing larger units. India has almost completed construction of its first breeder reactor, and their government expects to build more to utilize their large reserves of thorium, which their reactor can convert to uranium. China has a 25-megawatt prototype breeder reactor, with plans to build larger units.[9]

So how does nuclear power answer our primary questions about future energy sources?

- Is it free of carbon dioxide emissions? As of today, not entirely. While an operating nuclear power plant does not emit carbon dioxide, many support activities do. Construction of the power plant, mining and enrichment of the fuel, disposing of the waste, and decommissioning of the power plant all consume fossil fuels and contribute carbon dioxide. Those activities would all have to be converted to electricity, so they could operate on nuclear electricity to be carbon-free. Even then, under a nuclear growth plan of sufficient size to displace fossil fuels, the energy required for new plants would consume a substantial portion of the electricity from the old plants, pushing a net reduction of carbon emissions further into the future.[10]

- Is it big enough to replace fossil fuels? Only if breeder reactor programs replace the current nuclear fuel cycle.

Despite these limitations, many national governments were reviving dormant nuclear energy programs to meet their carbon-reduction goals. These programs went dormant for two reasons: Costs that were higher than the fossil-fuel alternatives, and public resistance to all things nuclear. When rising fossil-fuel demand drove energy prices higher, nuclear was becoming more competitive. And the public resistance barrier was generally viewed by nuclear proponents as ignorant fear that could be overcome with the right marketing message. The time was right for marketing nuclear as green energy to fight global warming. A nuclear renaissance was emerging.

Then came Fukushima.

Fukushima

In March 2011, a tsunami triggered by a major earthquake knocked out the cooling systems for three nuclear reactors in Fukushima, Japan. Without cooling, the reactors overheated, melted their cores, exploded their housing buildings, and spewed great quantities of radiation into the air and water.

Residents within a 20-kilometer radius of the plant were immediately evacuated, and that zone was later extended to 30 km. That area is likely to remain a "forbidden zone" for decades.

Japanese officials have been chasing the escaped radiation ever since the accident. Some examples:

- Farmers are stuck with 7,000 tons of radioactive rice straw contaminated with cesium. Some tried to feed it to beef cattle, but that led to a recall of all beef from the region for its radioactive contamination.[11]

- Tons of radioactive surface soil and vegetation must be removed from Fukushima and four adjacent prefectures. The estimated volume of 29 million cubic meters could fill the Tokyo Dome twenty-three times. Finding sites to dispose of the soil is proving difficult.[12]

- Three years after the accident, groundwater passing through the contaminated site continues to carry radioactivity into the ocean. Plant operators are trying to stop the water flow by burying a network of refrigerant tubes to freeze the earth into an underground "ice wall" ninety feet deep.[13]

The world has now witnessed three major nuclear power plant accidents over a three-decade period (Three Mile Island in 1979, Chernobyl in 1986, and Fukushima in 2011). The Fukushima "strike three" is driving many countries to abandon their nuclear plans. Germany will phase out all existing plants, and Switzerland will not replace their nuclear plants. When nuclear power was put to a public vote in Italy in June 2011, fully 94 percent voted against the government's plans to revive Italy's nuclear industry.

Yet over time, deteriorating climate will likely drive demands for greater carbon dioxide cuts, and such demands could overcome resistance to nuclear power. But nuclear power would still face the critical unsolved problem that has existed since its inception: radioactive waste.

Nuclear Waste

The fuel rods in a reactor must be periodically replaced to supply fresh nuclear fuel. The spent rods that are removed are far more radioactive than the ingoing fuel. In fact, the spent fuel could be classified as one of the most dangerous materials in the world. The danger comes from the jumbled mix of elements that generate many different kinds of radiation, as well as the overall intensity of the radiation. And each reactor routinely produces about thirty tons of such material every year.

The nuclear waste problem is not easy to fix, because it derives from the random nature of fissioning uranium. There are about twenty different ways that an individual uranium-236 nucleus can split. For example, one uranium-236 atom might split into a krypton atom and a barium atom, while another produces a strontium atom and a xenon atom. The path that any single uranium-236 nucleus takes is random. There is no known way to control the reaction to produce only certain fission products. So as the reactor operates, the original uranium-235 fuel is gradually transformed into a whole range of new elements.

Why is this a problem? Because some of the fragment atoms are themselves radioactive. For example, one split might produce a radioactive strontium-90 atom, while another might produce a non-radioactive strontium-88 atom. Even common elements like iodine, silver, and tin are generated in radioactive forms.

Some of these radioactive products further decay into other new elements, certain of which are also radioactive isotopes. A strontium-90 atom will decay into yttrium-90, which is also radioactive and decays into zirconium-90, which is not radioactive.

So the original fission process and the subsequent radioactive decay paths of the fission fragments result in a mixture of both benign and highly radioactive elements, none of which were present when the fuel was loaded into the reactor. The mix constantly changes and becomes more radioactive while under operation.

In the reactor, the radioactivity is shielded and contributes to the heat of the reactor, but the fuel rods cannot stay in the reactor forever. Some of the newly created elements absorb neutrons (without fissioning). If too many neutrons are absorbed, then the chain reaction cannot be sustained. At that point, the nuclear fuel has been poisoned by its waste products, and must be replaced with fresh fuel.

If nuclear engineers could control each fissioning atom to produce only non-radioactive fragment elements, then we would not have a nuclear waste problem. We would be able to operate a reactor to generate electricity and be left with a benign waste product that is easy to dispose of. The mix of newly generated rare elements might even make the waste valuable.

But such control is not possible. Just as burning fossil fuels always produces carbon dioxide, fissioning uranium-235 always produces radioactive waste. Once it comes out of the reactor, that waste must be handled with extreme care, because radioactive materials are deadly to living beings. The energetic particles they emit rip through living tissue like submicroscopic bullets, disrupting cells and tearing DNA. Thousands who survived the nuclear blasts at Hiroshima and Nagasaki later succumbed to radiation poisoning or radiation-induced cancer, and many of the women survivors later gave birth to deformed babies.

Even small doses can be fatal. In 2006, former Russian agent Alexander Litvinenko drank a cup of tea that had been surreptitiously laced with a few micrograms of radioactive polonium-210. He died three weeks later from radiation poisoning.

The spent fuel rods pulled from the reactor are so radioactively hot that they cannot be approached, and a robotic arm is used to immediately submerge the rods in onsite pools of water to shield the radiation and cool them. The rods are still generating copious heat because intense radioactive reactions are still taking place in them. They will continue to seethe with radioactivity long after the reactor has completed its life and shut down.

Radioactive half-life

Radioactive materials never disappear entirely once they are created. Each element decays at its own rate, which we measure as the *half-life*. That's the amount of time it takes half the radioactive atoms to decay and shoot out their radiation. You might think that if we wait just two half-lives then it would all be gone, but that is not how it works. During each half-life, half of what is left after the previous half-life fires off. If you start with four pounds of radioactive cesium 137, which has a half life of thirty years, during the first half-life two pounds will decay and you will have two pounds of radioactive material left. During the second thirty-year period, half of what is left decays, leaving one pound radioactive. After another thirty years it goes to a half pound, then a quarter pound, and so on. It's like two polite people sharing a dessert. Neither wants to take the last bite, so for each bite they take half of what is left. At that rate they will never finish the dessert. Likewise, radioactive decay never goes away entirely, it just gets small enough to be ignored relative to the general background radioactivity on our planet.

The levels of heat and radioactivity do decline over time. Many of the waste elements have short lives and decay to stable elements. Within five years, the overall radioactivity is less than 1 percent of what came out of the reactor. You might think, at that rate it wouldn't take long for it all to decay and disappear, but it's not that simple.

While the short-lived elements are disappearing, many long-lived elements are left behind, and many are still being *created* in the waste fuel. Take americium-241, for example. At discharge, there are 80 curies[14] of americium-241 per ton of spent fuel. After 180 days, there are 166 curies, more than doubled since being removed from the reactor.[15] That's because americium-241 is a product of the nuclear reactions that continue in the spent fuel after it is removed from the reactor. Americium-241 has a long half-life, 433 years, which means it will maintain appreciable radioactivity for thousands of years.

This is a serious problem with spent fuel—the longer it sits unprocessed, the more of these long-lived radioactive elements are created. Some are extremely long lived. For example, radioactive iodine-129 has a half-life of 16 million years, and it continues to be created in the spent fuel. A half-life of 16 million years means it essentially lasts forever. Iodine-129 is a particularly bad waste problem because it is soluble in water, and if ingested it is stored in the thyroid gland where it can cause thyroid cancer. In this respect, nuclear waste becomes more dangerous the longer it sits.

And because radioactive iodine-129 is chemically no different from common iodine, you wouldn't be able to tell you have absorbed it. That's why people spooked by radioactivity. It is invisible and deadly.

Radioactivity is especially insidious because it cannot be detected by any of your senses. If it gets loose, you cannot tell it's near you unless you have a Geiger counter. If you happen to come across some radioactive material, you won't know it because you can't see the radioactivity and you can't feel it. Yet the invisible radioactivity is attacking your molecules, damaging and disrupting your body's internal systems.

An angry lab technician at the University of California at Irvine in Santa Ana California was arrested in 1999 for assault with a deadly weapon for smearing a radioactive compound on a coworker's chair. The coworker sat on the chair for about six hours before a routine Geiger counter sweep revealed the contamination.

You can get a heavy dose of radiation exposure very quickly, even in a fraction of a second, if you are exposed to a strong source. Workers trying to clean up the Fukushima reactors can work for only a few minutes in the worst areas, and then they must not return for at least a year.

You can also get a large dose from a small source if exposed to it for a longer time, because the dosage accumulates over time. That can happen if you inhale a small particle of radioactive dust and it lodges in your lung. And radiation exposure of either type is usually irreversible. There is no cure for radiation sickness, and it is a horrible way to die.

The greatest tragedies hit those people who innocently handle radioactive material without knowing it. In one incident, a cylinder containing a strange glowing material was found in a Brazilian junkyard, taken home, and shared among friends and family members. It turned out it was from an abandoned hospital's radiation treatment machine that contained a pile of highly radioactive cesium-137. Four people died of radiation sickness, including a six-year-old girl.[16]

Of course, that radioactive material wasn't supposed to be in the junkyard, but something went wrong somewhere and it ended up there. No one knew until the people started dying.

The official line is that all radioactive materials are to be tracked from cradle to grave. But there's a big lie at the end of that official line—there is no grave. No facilities exist for long-term storage of high-level radioactive waste anywhere in the world. Currently almost all of the spent fuel is stored at the site of each reactor. This was supposed to be a temporary arrangement, but much of it has been sitting there a long time. Storing it in a more permanent place has been delayed because it has proven so difficult to devise a storage system.

The problem is daunting. The radioactivity from spent fuel rods is so intense that it must be carefully shielded at all steps of handling, repackaging, and transport. The material is also hot in temperature because the radioactivity generates heat. If it isn't constantly cooled, it will melt of its own generated heat. Because many of the elements remain radioactive for thousands of years, that is how long it must be kept isolated.

The US government has tried for decades to develop a permanent nuclear waste dump. The program has been beset by technical problems, political resistance, and changing policies. The main technical problem is the difficulty of predicting geological behavior over thousands of years. That's never been done, and is impossible to test. The political resistance comes mostly from states that are candidates for siting the facility. They object to becoming the dumping ground for the entire country's nuclear garbage.

The changing policies are simply the result of the program spanning eleven different US presidential administrations, from Eisenhower to Obama. The program's history is full of changes of direction and abandoned efforts. For example, the original plan called for the reactor wastes to be reprocessed to separate out reusable nuclear fuel and concentrate the deadly wastes. That proved to be too expensive and dangerous, and now the government just wants to store away the used fuel rods themselves. Of course, the rods were not designed to be stored for thousands of years.

Then there is the cost of developing the nuclear-waste dump site, for which the government would like the utilities to pay and the utilities would like the government to pay. It wasn't until 1982 that the government finally settled the issue and started collecting money from the utilities for waste storage. They collect a tenth of a penny per kilowatt-hour of nuclear electricity, which they put into the Nuclear Waste Fund. The idea was to investigate several possible sites and pick the best as a result of testing. Budget cuts reduced that effort to investigating a single site, with all hope riding on that site.

The government settled on Yucca Mountain, Nevada, as the site to bury the waste. They chose dry Nevada because over time water would corrode the storage canisters and release the radioactivity. If radioactivity gets into the groundwater, it could migrate into wells that people drink from. They chose Yucca Mountain because it is composed of tightly compressed volcanic ash that was supposed to prevent water flow.

But test tunnels at Yucca Mountain have shown some problems. The supposedly watertight rock turns out to be riddled with cracks that would permit water to flow. Even worse, they found chlorine-36 in the tunnel. Chlorine-36 is an element left on the ground by atomic bombs tested in the atmosphere in the 1950s. The chlorine-36 was carried 800 feet down to the repository level by rainwater percolating through the cracks in the rock. This discovery alarmed scientists because that meant it took only fifty years for water to reach the storage level, yet the storage site needs to stay dry for thousands of years.

They got another unwelcome surprise from the nearby Nevada Test Site where warheads were tested underground. It turns out that microscopic specks of clay contaminated with radioactive plutonium could actually float along with groundwater. The radioactive clay had migrated nearly a mile in just thirty years, much faster than they expected. That's half a foot per day, or about fifteen feet every month.

The Yucca Mountain project was twenty years behind schedule when the Obama administration pulled funding for it in 2009, after $12 billion had been spent on it. The United States is essentially starting over again.

Worldwide, thirty countries operate nuclear reactors, and thirteen of them have programs of some type to develop long-term waste storage. None of them have an operating facility yet. Only Finland and Sweden have chosen specific sites. France has an underground laboratory near Bure to study deep storage. All programs face technical and political uncertainties.[17]

The uncertainty derives from the unprecedented time frames during which the waste must be safely stored. The radioactive elements with the highest concentrations in spent fuel will take at least 10,000 years to decay—and 10,000 years is a *minimum*. Some of the radioactive elements in smaller concentrations remain dangerous for millions of years.

Deep geological storage could provide protection from the radiation, but no one knows for certain how the geology will behave for thousands of years. So we cannot just bury it and forget it, because the waste could leak into groundwater and eventually reach the surface to expose humans and other living beings. Storage sites will need to be maintained and monitored for leaks.

The nuclear programs have not budgeted for 10,000 years of managed care for nuclear waste, because it staggers the imagination to do so. The minimum maintenance period is longer than all recorded human civilization. How can we expect a government agency to remain stable and operating for 10,000 years?

Imagine if the ancient Egyptians had discovered nuclear power 5,000 years ago. At the time, they had no reason to believe that their civilization would not last indefinitely, so they would have carefully buried their nuclear waste and posted procedures written in hieroglyphs for checking on it and keeping people away from the dangerous material. Today, we know that ancient Egypt's civilization did decline and everyone forgot how to read hieroglyphs. Pity the pre-Rosetta Stone archaeologists digging in Egypt and unable to read the warnings.

These long time frames contrast with the short duration of the energy flow. A nuclear reactor creates electricity that flows and gets used up within seconds, but the nuclear process cooks up elements that emit dangerous radiation for thousands of years. The energy is transitory but the danger is essentially permanent. Each kilowatt-hour provides a fleeting benefit for today's energy consumer, while saddling that consumer's descendants with the responsibility of guarding the waste for 400 generations or more.

Yet all those future generations get no benefit from the nuclear power that created this waste. We are the generation that used up the electricity running our toasters and air conditioners. The useful energy is long gone, but the radioactive waste persists. All they get is the burden of watching over our dangerous waste. That is grossly unfair to all those future generations.

Responsible people have proposed a moratorium on constructing new nuclear power plants until nuclear waste repositories are operating. Nine states in the U.S. have already enacted such bans[18] in response to fifty years of broken promises about solving the nuclear waste problem.

Thorium Reactors

There exists another form of nuclear technology that might solve many of the current problems with nuclear power—thorium-fueled reactors. The element thorium is itself not radioactive and cannot produce power. But if a thorium atom absorbs a neutron, it becomes uranium-233, which can fission to produce energy.

By mixing thorium with its uranium fuel, a nuclear reactor can generate power from the uranium and at the same time breed more fuel from the thorium. After a period of such operation, the fuel rods are removed and the newly bred uranium-233 is separated out and made into new fuel rods.

The potential advantages of a thorium-based nuclear fuel cycle include:[19]

- Thorium is far more abundant than fissionable uranium.

- The nuclear waste from a thorium reactor cools in hundreds rather than thousands of years.
- New reactor designs could be immune to meltdown accidents.

The potential disadvantages include:

- The spent fuel must be reprocessed to make new fuel rods, and the feasibility of thorium reprocessing has yet to be proven.[20]
- Thorium fuel cannot be used in existing reactors,[21], so new reactor designs must be developed, tested, and analyzed before commercial deployment.
- The uranium-233 produced from thorium can be made into a nuclear bomb, increasing the risk of nuclear-weapon proliferation.[22]

All of these advantages and disadvantages are "potential" because the development of thorium reactors is still in its early stages, even though they were demonstrated in the 1960s.

A fateful decision in 1973 turned the thorium fuel cycle into the neglected stepchild of the nuclear industry. The US government, which had been pursuing research in both uranium- and thorium-fueled reactors, decided to shut down all thorium research in favor of uranium-fueled reactors.

Today, thorium reactors are being "rediscovered," with India as the leading proponent. India has extensive thorium reserves, and has embarked on a multistage development plan.[23] However, full exploitation of India's thorium reserves is expected to take several decades.[24]

Fusion Reactors

Nuclear proponents tell us about another form of nuclear technology that does not generate vast quantities of nuclear waste—fusion power. That's true, a fusion reactor is very different. In a fusion reaction, hydrogen atoms are fused together to make helium atoms. Only atoms on the light end of the element table can fuse, and when they do, the energy they release is proportionally much greater.

This is a much cleaner reaction because it isn't smashing fat atoms into random radioactive fragments as in a fission reactor. Most people know the end product, helium, as an inert gas, safe enough to inflate kids' balloons with. The hydrogen-to-helium reactor also yields more energy per unit of fuel, which is why hydrogen bombs are so much more powerful than uranium bombs.

But fusion has proven extremely difficult to sustain in a controlled reactor. Fusion reactions tend to fly apart from their own energy. They must be contained with a very powerful force, or the reaction stops. Fifty years and billions of dollars of research have not found a way to contain a fusion reaction for practical power generation. That research gamble has not paid off like other high-tech ventures have.

Even if fusion reactors did work, they would still produce some radioactive waste. The structure used to contain the fusion reaction would be bombarded by neutrons. Through neutron activation it would become radioactive over time. That means fusion reactors would still create radioactive waste, just a lot less of it.

There is one approach to nuclear energy that might prove acceptable to the public—put a fusion reactor out in space. This sounds far-fetched, but could work. The fusion reactor would have to be far enough away so that if there were an accident, the radioactive mess would not drift down to earth. And we don't want to rocket any dangerous materials up there because rockets do explode on occasion (remember the Challenger and Columbia Space Shuttle explosions). So we should assemble the reactor entirely in space from material already there. Since it is out in space, we can make it really big, because we aren't supporting weight or handling earthquakes. We could keep growing it to a huge size. In fact, if it gets big enough, the force of gravity will hold the fusion reaction together so it can sustain itself. That would eliminate the problem of the container becoming radioactive from neutron activation. Of course, we need to solve the problem of getting the energy down to earth. We would probably beam it down in the form of electromagnetic radiation, which carries energy quickly and efficiently through the vacuum of space. We would have to select frequencies that easily pass through the atmosphere and are not harmful, and it would be best to distribute the energy over the surface of the earth so subscribers could pick it up directly.

Fortunately for the human race, that future has already arrived. We have such a fusion reactor today—the sun. If the sun were built by engineers, it would be hailed as a brilliant design for a fusion reactor.

The sun certainly is a fusion reactor, steadily burning its hydrogen in a nuclear reaction to make helium in its core. It is entirely self-contained by its own gravity, which is a great force for containment because it is so constant, and therefore so reliable. The Sun-as-fusion-reactor is completely self-regulating, so it has no operating costs and no maintenance costs. It is built to last, about 5 billion years. It puts out far more energy than humans use. It delivers its energy in a benign form, sunlight, and it distributes that energy all over the Earth's surface for anyone to pick up. It generates no carbon dioxide to add to climate change, no air pollution to cloud our cities, and no nuclear waste for future generations to guard. With specifications like those, the sun is the fusion reactor that all others must aspire to.

In other words, why spend billions of dollars trying to duplicate the sun's fusion reaction on Earth, when we can just make use of the fusion reactor we already have?

PART II: OUR ENERGY DILEMMA

Solar is The One

CHAPTER 12

Solar Energy

O ur giant fusion reactor in the sky bathes the Earth in a continuous supply of high-quality solar energy, spreading it out over all parts of the planet for anyone to gather and use. The sun has been performing this service for millions of years, so why have we made so little use of this energy source?

The problem is not that we are unfamiliar with solar energy. Your ability to see during daylight hours is enabled by solar energy. When you park your car outside on a sunny day with the windows rolled up, you directly experience the power of solar energy to generate heat.

The problem is not that solar energy is rare. Unlike oil and coal deposits, which are found in only a few places in the world, solar energy is *distributed energy*. It is available in all countries for anyone to put to use.

The problem is not that the amount of energy is too small to bother with. A section of roof twenty feet by thirty feet located in Sacramento, California, receives solar energy equivalent to over 3,000 gallons of gasoline each year.[1] For comparison, the average car in America consumes only 530 gallons per year.[2]

The fundamental problem with solar energy is that it is different. In almost every way imaginable, solar energy differs from the fossil-fuel energy we depend on today. Fossil fuels are dug or pumped out of the ground by energy companies, who sell the energy as a commodity, either directly in the form of motor fuel like gasoline, or indirectly in the form of electricity. In contrast, solar energy falls on our heads from the sky, arriving directly on site without the aid of any energy company.

Because solar energy arrives directly to you from the sun, the energy is free. Of course, the equipment to harvest it is not free, but once you install the equipment, you do not have to pay a monthly bill for the energy it produces. This difference complicates economic comparisons with fossil fuels, which are sold by volume.

Solar is also a fundamentally different form of energy. If solar energy rained down on us in the form of a liquid like gasoline, we could just put out buckets to collect it, pour it into our car's gas tank, and drive off. But solar energy arrives in the form of light, an energy form so unique that it takes some getting used to.

For example, unlike most other forms of energy, light is not attached to matter. Chemical energy is stored in the molecular bonds of matter, gravity energy is elevated matter, kinetic energy is moving matter, and heat energy is matter in motion at the molecular level.

Light energy is pure energy, free of matter. That's why solar energy does not pollute. There are no chemicals attached to sunlight to leave behind as harmful residues. And that's why solar energy does not generate carbon dioxide like fossil fuels, because it arrives with no carbon attached to it.

Being free of matter, light energy can easily traverse the vacuum of space to reach us from the sun. Light, which is the fastest thing in our universe, makes the journey in just over eight minutes, traveling at 186,000 miles per second (about a million times faster than sound).

What is light?

Light energy is composed of a tiny force field of electricity and a tiny force field of magnetism, both traveling at the speed of light. The electric field oscillates between positive and negative, and the magnetic field oscillates in synchrony between north and south. The rate at which they oscillate determines the frequency of the light.

The great speed of light presents one of the difficulties of working with solar energy. Light cannot be slowed down, stopped, and stored in its original form. Light must always travel at the speed of light, so you cannot just put light energy in a bottle and store it; you have to convert it to a form of energy that is stationary.

Fortunately, sunlight is a high-quality form of energy that can easily be converted to other forms of energy. Plants have been converting sunlight to chemical energy through photosynthesis for millions of years. Reptiles basking in the sun have been converting sunlight into heat energy for almost as long.

Humans have also been putting photosynthesis to work for thousands of years through the agricultural systems that produce all of our food energy. And in the early 1900s we learned to produce useful heat from sunlight using solar water-heating panels, which are now common in many parts of the world.

But the conversion of solar energy that will make it important as a modern energy source uses an entirely new process that is unlike anything in nature—a process invented by human minds and created with human hands. That process is the direct conversion of sunlight to electricity through photovoltaic panels.

Solar panels made from photovoltaic (PV) materials are modern miracles of energy conversion. They just sit in the sun and generate electricity, which can be directly substituted for the fossil-fuel-generated electricity coming from the power company.

The key to PV panels is the special materials they are made from. The most common type of panel today is made from silicon, essentially recrystallized sand. What makes them photoactive are additional elements infused into the surface that allow the crystal to interact with light, converting the light to electricity.

Home Solar

Making photovoltaic panels requires our most advanced high technology, but putting them to use is so elegantly simple that anyone can do it. Because sunlight is a distributed energy source, anyone can produce their own electricity simply by bolting PV panels in a sunny location and running wires to make the connections.

I know this for a fact because my wife and I have been using solar PV as our primary source of electricity since April 28, 1997. We installed PV panels on our garage roof when we built our house, and that was the

Figure 25. Photovoltaic panel

When exposed to sunlight, a photovoltaic panel produces flowing electricity, with no moving parts, no noise, and no undesirable byproducts.

date we turned the system on. We have been living on solar electricity since that day.

The PV panels connect to a control panel that manages the power from the panels. When electricity is needed, the PV power is routed to an *inverter*, an electronic box that converts the PV direct current (DC) to standard alternating current (AC) for use by our lights, appliances, computers, and tools.

You might think we live a primitive lifestyle to survive on solar electricity. In fact, our home is a normal modern American home, with a full complement of appliances, electronics, tools, and labor-saving devi-

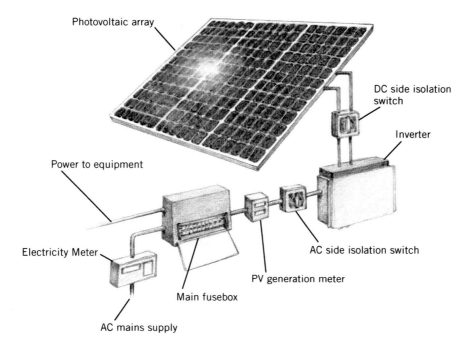

Photovoltaic array

DC side isolation switch

Inverter

Power to equipment

Electricity Meter

AC side isolation switch

PV generation meter

Main fusebox

AC mains supply

Figure 26. Photovoltaic system

A typical photovoltaic system consists of the panels to generate electricity, an inverter to condition the power, and the necessary switches and electronics to safely connect the system to the distribution box and power grid.

ces. Our solar electricity is of high quality and fully compatible with all of our electronic equipment. Friends and family notice nothing out of the ordinary when they visit.

The main difference is that most of our "stuff" is more energy efficient than most. We took care when choosing electrical devices that they be the most efficient we could afford. For example, we use compact fluorescent and LED lights instead of incandescent bulbs everywhere. These new lights use one-quarter of the electricity for the same amount of light, and they last much longer. We also purchased a high-efficiency refrigerator, dishwasher, and clothes washer.

You might wonder what we do when it's cloudy, or at night when the sun doesn't shine? That problem is readily solved with our battery energy-storage system, which matches supply with demand.

When the PV system generates more electricity than we are presently using, the excess is routed to a bank of batteries for storage. At night and on cloudy days we draw electricity from the batteries. We need battery storage because our home is not connected to the electrical grid like most homes. During winter we sometimes experience many days of cloudy weather that deplete the batteries, so we also have an electrical generator that runs on gasoline as a backup power source.

During spring, summer, and fall, we do not run the generator at all. Since we are not connected to the grid, that's positive proof that all of our electricity during those seasons comes from our PV system. We have three seasons of carbon-free energy, and one season of mostly carbon-free energy. Overall, about 90 percent of our electricity comes from the Sun.

We are always on the lookout for more ways to substitute solar energy for fossil fuels. Although we have a gas stove, we reduce gas for cooking by using a microwave oven, a small electric convection oven, an electric tea kettle, and an electric grill. These appliances draw a high current, but they don't run for very long so the electricity use is moderate. On sunny days, we can use our solar oven, a simple insulated box with a glass cover. I just aim it at the sun and it cooks my bread.

For stovetop cooking, we can replace a gas burner with an electric *induction cooker*. An induction cooker has a glass surface with magnetic coils underneath. It uses electricity to generate an oscillating magnetic field. The magnetic pan bottom absorbs the magnetic field, and converts the energy to heat directly in the pan metal. Induction cookers can heat a pan as quickly as a gas burner, and cool off quickly because the burner itself does not need to get hot. Induction cooktops are also more efficient than regular electric burners because all the electrical energy ends up as heat in the pan, not in the cooktop or the air around the pan.

For our yard, we have an electric hedge trimmer, electric chainsaw, and electric lawnmower. Gas-powered lawnmowers are notorious for the large amount of air pollution they produce for their size. Our solar-powered electric lawnmower produces no air pollution, and is quieter too.

In addition to our solar electricity, we use direct solar heat to further reduce our use of fossil fuels. We have a solar water-heating system to supply hot water, with a propane backup water heater if there's not

Figure 27. Sun oven

Sun ovens are having a big impact in Africa. A sun oven replaces firewood, which has been overharvested in many areas and requires women and children to spend hours gathering from miles away. The nonprofit Solar Cookers International has enabled over 30,000 families in Africa to use solar ovens, freeing up their time for more productive activities (including education), and reducing the ecological degradation from overharvesting firewood.

enough sun. There are millions of solar water heaters in the world today, with 1.5 million in Tokyo alone.[3] You can even buy panels from SunEdison or Solimpeks that combine PV and solar water heating in a single panel.

We also partially heat our home with the sun's energy. Our house uses a passive solar design. Here *passive* means there are no active fans or pumps to move heat around. The house is designed to admit winter sun through south-facing windows, and to retain the heat with insulation. Concrete walls absorb and release the heat to reduce temperature swings. In summer, we cool the concrete walls by circulating night air, and the cool walls keep the house as comfortable as an air conditioner would during the day. If that is not sufficient, we run a standard air conditioner off our solar panels. Solar PV is a good match for air conditioning because the hottest days are also generally the sunniest days.

Heating and cooling of the house could also be accomplished by feeding solar electricity to a *heat pump*. A heat pump is like a heat engine in reverse—instead of using heat to make mechanical energy, it uses mechanical energy to make heat. Instead of just generating the heat through friction, it pumps existing heat from the air or ground outside of the house to the inside.

A heat pump reverses the normal flow of heat energy. Heat naturally flows from hot to cold, but a heat pump forces heat to flow from cold to hot. Your refrigerator is a heat pump, using electricity in the compressor to pump heat from the cold interior of the refrigerator out into the warmer room.

With a heat pump, you can extract three joules of heat energy for each joule of electricity. That sounds like a 300 percent efficiency, but it does not violate the Principle of Conservation of Energy. Heat energy is not being created, but is just being moved from the soil to the house through the actions of the heat pump. Heat pumps can be reversed to supply air conditioning during hot weather.

We don't currently use a ground-source heat pump because it consumes more electricity than we can produce with our existing solar panels. But solar panels are modular, so we can easily add more panels when we can afford them. Even with more panels, though, our batteries can only store so much electricity, and a ground-source heat pump in our climate runs the most in winter, when we have the least solar energy available.

Most people with solar electric systems are connected to the utility grid, so they don't need batteries. When they have excess solar electricity, it feeds back into the grid and runs their meter backwards. Other utility customers connected to the grid use that solar electricity when it flows. At night and on cloudy days the solar owner draws power from the grid like other customers.

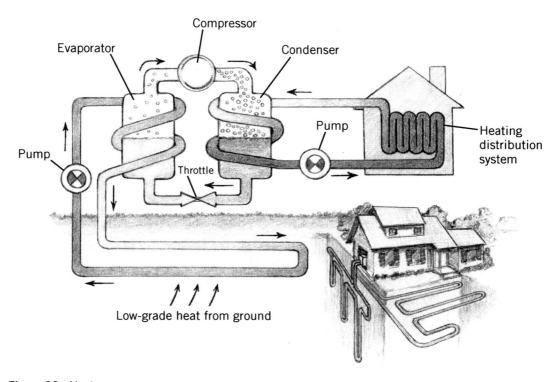

Figure 28. Heat pump

A ground-source heat pump circulates water through pipes buried in the ground to pick up the tepid heat in the soil. The compressor then raises the temperature of that heat for use in the house, for heating rooms and hot water. The heat pump in effect refrigerates the soil and transports the heat it extracts into the house. Heat then flows from the warmer surrounding soil to replenish the lost heat. The heat energy resident in the soil comes primarily from solar energy falling from the sky, not from the center of the Earth. Ground-source heat pumps are thus another application of solar energy.

In this sense they are using the grid to bank solar electricity, like a regular bank is used to store money. When you feed excess solar electricity into the grid during the day, you make a deposit. When you pull electricity from the grid at night, you make a withdrawal. Your solar electricity is not physically stored, but is consumed by someone else to offset their fossil fuel use. Your solar credit is stored until you draw electricity from the grid.

Most grid-connected solar electric customers subscribe to *net metering* with their utility company. In this scheme, if they use less electricity than they produce in a year, then they pay nothing except for a fixed connection fee. If they use more than they produce, they pay only for the difference. Most strive for a zero balance each year. Because the billing period is for a year instead of a month, they effectively bank summer electricity for use in winter, not unlike storing the excess food from a summer garden for winter meals.

With proper design and enough PV panels, you can build an *energy-positive* home, that is, a home that generates more energy than it uses. For example, eco-conscious developer Transformations Inc. has built dozens of energy-positive homes in Massachusetts. The homes have high levels of insulation, ground-source heat pumps, and large PV arrays integrated into the roof.

Solar Energy for Cars

Solar PV can also revolutionize personal transportation. Before electric cars became available from car manufacturers, I acquired a 1980 Volkswagen Rabbit that had been converted to electric power. The engine had been replaced with an electric motor, and the gas tank by a bank of batteries. When we charged the batteries from our solar panels, we drove a solar-powered car that produced no air pollution and no carbon dioxide.

Car manufacturers like Ford and Nissan are finally making electric cars, so you won't have to convert your own. But switching everyone to electric cars is not enough. Zoom out your energy scope to wide angle and follow the electricity to its source. If electric cars are charged with conventional electricity, then they are still largely powered by fossil fuels because most electricity is still generated by fossil fuels. Instead of air pollution and carbon dioxide coming out your tailpipe, they come out the stack of the utility power plant. An electric car charged from solar panels breaks that link to fossil fuels.

The big advantage of electric cars is that they give us the *opportunity* to run a car on solar energy by charging it up with PV panels. An electric car gives you choices of energy sources, while a gasoline car can only be powered with gasoline. I usually charge my electric car at home from my solar panels, but sometimes I park in our town's parking garage and connect to one of their charging stations. That bit of conventional electricity in our region comes from a mix of natural gas, hydroelectric, and nuclear energy sources. Until the town arranges to use only "green" electricity from solar, wind, and hydro for the charging station, I must compromise a bit.

The biggest problem holding back electric cars is *range anxiety*, the fear of exceeding the range of your batteries before you can reach a charging station. No one wants to be stuck on the roadside with depleted batteries and no place to plug in. The other major problem is the long time it takes to recharge once you do find a plug.

Tesla Motors Inc. has attacked these problems directly with a patented Supercharger system. A Supercharger can provide a 50 percent fillup in twenty minutes. Tesla has installed dozens of Supercharging stations in North America, Europe, and Asia to support their cars. More recently, Tesla freed up its Supercharger patents to make it the industry standard for all electric cars. Tesla also built battery-swapping stations that can swap in a set of charged batteries in ninety seconds, faster than a typical gasoline fillup.

Charging issues have also spurred development of *plug-in hybrid* cars, which merge an electric car with a gasoline car to negate each other's problems. It differs from a conventional hybrid by carrying a larger battery bank that can be plugged in for charging while the car is parked. The car can run on the electric charge until it is depleted, and then switch over to the gas engine to complete the trip, thereby eliminating range anxiety. If the trip distance is within the range of the batteries, the gas engine need not come on at all. No need for two cars—a single car can run entirely on electricity as an electric car for short trips around town, and burn gas for long road trips.

If the charging electricity comes from solar panels, then the car is solar powered while running on batteries. The compromise comes on longer trips that deplete the battery. Then the gasoline engine kicks in and you are driving a conventional car.

Plug-in hybrids can be viewed as a transition car—a bridge to future electric cars whose batteries can match a tank of gas in terms of distance and charging time. Or the gasoline engine might be replaced with a fuel cell that runs on a solar-generated fuel such as hydrogen or methanol. More on fuel cells later.

Electric cars (including plug-in hybrids in electric mode) are more efficient than gasoline cars in two other ways. An electric car does not have to idle its motor when stopped in traffic. A gasoline engine must idle to keep it available for startup, but an electric motor can easily start from a dead stop. In city traffic, idling wastes a significant portion of the gasoline energy.

Electric cars also tap into a new energy "source" by scavenging the car's own kinetic energy. Most electric cars have what is called *regenerative braking*. When the brake pedal is pressed lightly, the car's brake pads do not move. Instead, the electric motor switches mode and becomes a generator, which converts the kinetic energy of the car into electricity that charges the car's batteries. Draining the kinetic energy slows the car, just as brakes do, but instead of simply throwing the energy away as heat in conventional friction brakes, regenerative braking recycles the car's kinetic energy. For safety, the friction brakes engage when you press harder on the brake pedal.

With regenerative braking, a car's braking energy system changes from an energy sink to an energy source. Such energy recycling is not possible with a gasoline car because there is no easy path to convert kinetic energy to gasoline. Given the number of cars on the road and the number of times they must brake on each trip, regenerative braking is a vast energy source that was not available before electric cars.

Solar for Business

Commercial businesses could be run almost entirely on solar electricity, generated either onsite or offsite. Businesses use electricity for lighting, air conditioning, and office equipment. Most businesses operate during the day, when the solar resource is available, so they can use solar electricity directly from PV panels.

Most commercial buildings have flat roofs that have no shading and cannot be seen from street level. Such roofs are ideal for installing solar PV panels. Retailers Walmart and Whole Foods have many megawatts of solar PV installed on their buildings. Schools, hospitals, and other institutions also typically have flat roofs that could hold PV panels.

Businesses such as restaurants and hotels that need hot water can add solar water-heating panels. These can be installed alongside the PV, or merged with the PV by using hybrid panels that produce both electricity and hot water.

For new buildings, solar PV components can be integrated into the structure as part of the building's external skin, rather than added on. Such *building-integrated PV* (BIPV) essentially disappears from view while being in plain sight. It can also save money. Instead of building a roof and then adding panels, the panels become the roof. In a 2009 demonstration project, Hydro Building Systems constructed a building in Bellenberg, Germany, that uses building-integrated PV to produce power and run heat pumps for heating rooms and water. This building has an innovative facade design that reduces its energy consumption. It exports 80 percent more electric power than it uses, making it an energy-positive building.[4]

Solar for Industry

Solar is fine for homes and small businesses, you might think, but what about industry? Adapting solar energy to industry requires a bit more creativity because each industry is different. Industry has a vast number of ways of using energy, but they all distill down to just a few basic forms of energy input: electricity, heat, and fuel. All of these can be produced from solar energy.

A car factory, for example, runs almost entirely on electricity to move the assembly line, animate robots, operate power tools, and light the work areas. While factory roofs are usually large and flat, they do not provide sufficient area for PV to power the entire plant. A factory can choose to import solar electricity or set up their own offsite arrays of panels. Volkswagen installed thirty-three acres of solar panels offsite to provide power to its Passat factory in Chattanooga, Tennessee.[5] But even that large system can only provide 12.5 percent of the factory's electricity needs, so they will still need to import solar electricity from offsite.

A food-processing plant uses electricity, but also needs large quantities of steam and hot water. The required water temperatures are higher than what can be supplied by flat solar water-heating panels, so *parabolic trough* systems can be used instead.

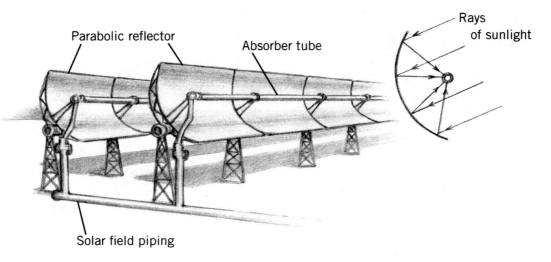

Parabolic reflector

Absorber tube

Rays of sunlight

Solar field piping

Figure 29. Parabolic trough solar collectors

Trough-shaped mirrored collectors with a parabolic cross-section tightly concentrate sunlight onto a black pipe running the length of the trough. The heat-transfer fluid circulated through the pipe can reach 750°F. The trough opening rotates to keep the sunlight focused on the pipe, tracking the sun as it moves across the sky.

When there is not enough sun, the heat can be supplied by stored heat, stored solar electricity, or fuels derived from solar energy.

Even steel foundries can go solar. One-third of the steel made today already substitutes electricity for coal in electric arc furnaces. Such plants are far more energy efficient than coal—it takes 20 gigajoules to make a ton of steel using coal, but only 7 using electricity.[6] Since steel foundries are the largest industrial emitters of carbon dioxide,[7] converting them to solar could have a significant impact.

Cement plants and some other manufacturers may still require high-temperature fire, so some form of fuel will still be needed. But such fuels can also be derived from solar energy, as described in the section titled "Solar Fuels" (page 150).

Most industrial plants consume more energy than they can produce on site from solar energy systems. They will need to rely on energy generated elsewhere, so there will still be a call for large solar-based power plants. Such big solar systems can come in many forms.

Big Solar

Rooftop solar PV systems are limited by the available roof area to a few kilowatts of power. Such limitations are removed in ground-mounted PV systems. Utility companies and private investors around the world are installing huge arrays of PV panels on acres of land dedicated to generating solar power. For example, the Agua Caliente Solar Project exposes 290 megawatts of PV panels to the blazing Arizona sun. The power from the 4 million PV panels flows into the power grid to supply customers of the Pacific Gas & Electric utility company in California.

Deploying fields of mirrors instead of fields of PV panels, solar thermal power plants generate electricity by concentrating sunlight onto a heat engine, as with this power plant in Spain.

Figure 30. Abengoa Solar PS20 concentrating solar power plant

The 1,255 pivoting mirrors of the Abengoa power plant in Spain are automatically steered to reflect sunlight onto a small area at the top of the central tower. The concentrated sunlight generates high-pressure steam at 300°C to drive a conventional steam turbine to produce 20 megawatts of electricity.

To achieve even larger outputs, Spanish engineers replaced the "power tower" concept in later plants with fields of simpler (and cheaper) parabolic trough collectors (see illustration on page 144). Each trough rotates to focus the sun onto a pipe carrying a thermal transfer oil, heating it to 400°C. The oil circulates to a central power plant where it boils water to drive a steam turbine. As of 2012, Spain had eleven parabolic systems of over 100 megawatts each.

The Andasol 1 Solar Power Station in Spain goes one step further. During the day some of the collected heat is transferred to a heat-storage system made of special molten salts. During cloudy periods or at night, the system recirculates the stored heat to generate power when the sun is not available. With about seven hours' worth of energy storage, the plant can supply electricity 70 percent of the time instead of the 25 percent it could achieve without energy storage.

Hydroelectricity

Most people are surprised to learn that the cheapest source of electricity today derives not from coal but from solar energy, indirectly through water power. Hydroelectricity is solar-based power, because the water is lifted by the sun's energy.

Electricity from hydropower dates back to the 1880s and is a well-proven technology that emits no carbon dioxide or other pollutants during operation. Hydroelectricity could be expanded to replace fossil-fuel electricity, but it often meets resistance because of the environmental impact of dams. Dams are usually constructed in mountainous areas, and may flood large areas of wildlife habitat or traditional agriculture. The Three Gorges Dam in China displaced 1.3 million people from their homes and livelihoods. Fish migration runs become blocked, and silt accumulates behind the dam instead of flowing downstream.

A dam is not a required part of a hydroelectric system. You can run a water turbine by diverting a stream flow at a higher elevation through a pipe that runs downhill directly to the water turbine. This is called a *run-of-the-river* hydroelectric system. For example, the Upper Mamquam Hydroelectric Plant in British Columbia delivers 25 megawatts of electricity and has no dam.

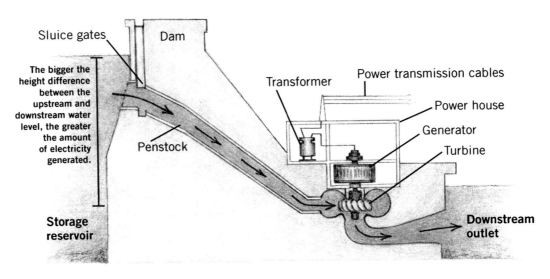

Sluice gates Dam

The bigger the
height difference
between the
upstream and
downstream water
level, the greater
the amount
of electricity
generated.

Penstock

Transformer

Power transmission cables

Power house

Generator

Turbine

Storage
reservoir

Downstream
outlet

Figure 31. Hydroelectric power
Hydroelectric systems use the power of falling water to drive turbines connected to electrical generators. Today 16 percent of the world's electricity comes from hydropower (more than nuclear energy).[8]

The Niagara Falls hydroelectric system also operates without a dam. Between 50 percent and 75 percent of the Niagara River's flow is diverted upstream of the falls through tunnels that flow to water turbines some distance below the falls. The elevation change is modest (about one-fourth of the height of Hoover Dam), but the huge water flows enable the system to generate up to 4.4 gigawatts of electrical power. During the Niagara Falls tourist season, diversion for hydropower is limited to 50 percent during daylight viewing hours to ensure a good showing of the waterfalls.

Most dams are not built solely to generate electricity but to control flooding and supply irrigation water for farming. They capture the peak river flows behind the dam to prevent flooding, and allow the stored water to be metered out during dry periods. Running the released water through a water turbine provides a bonus of electricity. The value of the electricity typically helps pay for the dam over time. Once the dam is paid off, the low maintenance costs and free solar energy make hydroelectricity the least expensive of all electricity sources today.

A dam makes it easy to manage hydroelectricity output by storing gravity energy behind the dam to convert on demand. A run-of-the-river hydroelectric system lacks the energy storage of a dam, so its output typically tracks the fluctuating water flow of the river.

Another type of dam-free hydroelectric power is being developed with turbines submersed in large flowing rivers. They operate like wind turbines but under water.

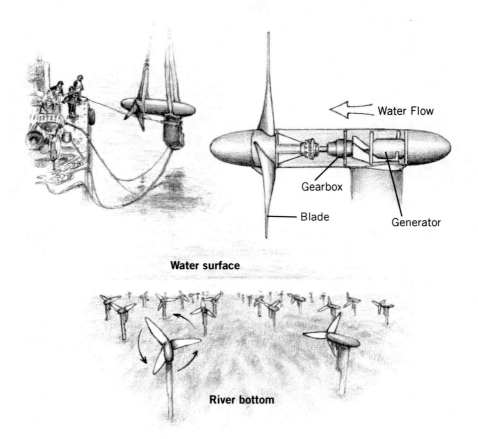

Figure 32. Hydropower in East River

Like a wind turbine under water, free-flow hydropower blades are turned slowly by the moving river, transforming a portion of the river's linear kinetic energy into rotational kinetic energy. The turbine's rotational energy is connected through a gearbox to an underwater generator, and power is brought up through cables to the shore.

Such *free-flow* systems rely on the direct kinetic energy of the flowing river water, and not on gravity energy, so no dam is needed. Verdant Power has demonstrated such turbines in the East River in New York City.

Wind Energy

Wind energy provides another way to generate large quantities of electricity indirectly from solar energy. Wind comes from solar heat creating high- and low-pressure zones that move large air masses across the planet's surface. The moving air contains kinetic energy that can be extracted by a wind turbine to make electricity. The extractable power of the world's winds is about sixty times greater than all of human energy use.[9]

Wind turbines are the second largest source of solar electricity today, behind hydroelectricity. Modern wind turbines use airfoil blades similar to airplane wings. Since the power is proportional to the area swept by the blades, doubling the length of the blades quadruples the power. That factor has pushed blade diameters up, with the largest being a 7-megawatt machine with blades sweeping a circle 413 feet in diameter. For comparison, the wingspan of a Boeing 747 is only 225 feet. Like hydroelectricity, wind turbines are a well-proven technology and are being commercially installed worldwide.

Unlike direct solar energy, wind energy is not evenly distributed. The power in the wind is proportional to the cube of the wind speed, so doubling the wind speed yields eight times more power. It pays to find a very windy location for a wind turbine, typically a mountain pass, a windswept plain, or the open ocean. When such places are found, it also pays to put up more than one turbine, so you will see wind farms with rows of wind turbines spinning.

The economic incentives for using larger turbines and multiple turbines in windy locations have turned wind energy into another form of Big Solar. The largest wind farm as of 2014 was the Gansu Wind Farm in China, with a combined peak generating capacity of over 5 gigawatts.[10]

Although wind-power installations generate no pollution or carbon dioxide during their operation, they have experienced some resistance regarding visual impact and bird kills.

One solution to the visual impacts is to place wind turbines offshore. Although the marine environment is more difficult for building and connecting wind turbines, the winds are generally stronger and steadier than on land. Denmark built the first offshore wind farm in 1991, and now Europe has over forty offshore wind farms with a combined capacity of 4,300 megawatts. The European Wind Energy Association projects the offshore capacity to grow to 150,000 megawatts by 2030.[11].

Bird kills are a more difficult problem to solve. Although large wind turbines seem to rotate at a leisurely pace (the largest take a full five seconds to complete one rotation), the tips of the blades are moving at six times the speed of the wind. In a 20 mph wind, the blade tips are traveling 120 mph, fast enough to catch many birds off guard.[12]

Studies have shown that bird deaths can be minimized by locating wind farms away from bird migration paths and by using a bird-friendly blade design. Actual bird counts have shown that the number of birds killed is relatively small. According to the U.S. Fish and Wildlife Service, 440,000 birds are killed in the United States by wind turbines each year.[13] A study published in 2013 in Nature Communications estimated that free-ranging domestic cats kill between 1.4 and 3.7 *billion* birds each year.[14]. We can save far more birds by controlling cats than by stopping wind farms, and we can even save human lives by replacing coal power plants with wind farms.

Solar Fuels

Certain energy uses will still require liquid fuels: Ships, airplanes, long distance trucks, and field equipment used on farms, such as tractors and combines. None of these can be tethered to an electric cord, nor can they realistically carry sufficient batteries. They need portable stored energy.

Liquid fuels such as gasoline and diesel fuel have two big advantages: they store a lot of energy in a small volume, and they are easy to pump for refueling.

Not all transportation needs liquid fuels. Many trains run on electricity delivered through the tracks, including high-speed bullet trains. These could be solar powered if their electricity came from a solar source such as PV or wind. A two-mile-long Belgian train tunnel along the Paris-Amsterdam high-speed rail line, which was built to shelter trains from falling trees, has now been topped with 16,000 PV panels. Known as the Solar Tunnel, its power is used by the trains and stations.[15]

Long-distance trucking can also be largely replaced by electric trains powered by solar electricity. The truck trailers can be efficiently piggybacked onto rail cars for long-distance travel. A liquid-fueled cab would be needed only for the final leg.

Electric cars do not need liquid fuels, but such fuels are needed by plug-in hybrid cars, which fire up an internal combustion engine when the battery runs out. If a plug-in hybrid could run on liquid solar fuel when its solar electricity ran out, then the car could be 100 percent solar powered. Similarly, a jet airplane could be solar powered if its fuel was derived from solar energy.

Is it possible to make liquid fuels from solar energy? Not only is it possible, it is relatively easy, using methods that have been in use for decades. The process has two steps: First you generate hydrogen gas by breaking down water using solar electricity, and then you combine the hydrogen with carbon dioxide to make liquid fuel.

The hydrogen can be generated by passing solar electricity through water, which splits the water molecules into hydrogen and oxygen gases, a process known as *electrolysis* (see the following illustration). The collected oxygen can be sold as a separate product to hospitals, and the hydrogen is passed to the second step. The carbon dioxide required for the second step could initially come from the exhaust stack of a fossil fuel power plant, but more advanced systems in the future could simply extract it from the air.

These gases are fed into a chemical reactor vessel, where a process invented by Franz Fischer and Hans Tropsch applies high temperatures and catalytic materials to trigger chemical reactions that combine the gases into a liquid fuel. By varying the catalyst material, temperature, and pressure, different fuels can be produced, including gasoline, diesel fuel, methanol, and dimethyl ether (similar to propane). Additional chemical steps can produce almost any other petrochemical product, including jet fuel.

The Fischer-Tropsch process has a long history. Invented in Germany in 1925, it was used to manufacture liquid fuel from coal in World War II Germany when that country was cut off from normal oil imports.

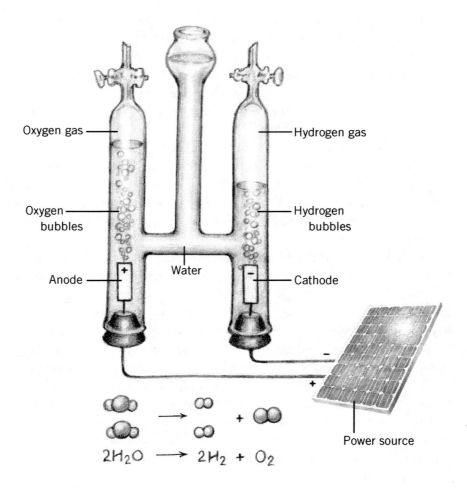

Oxygen gas

Hydrogen gas

Oxygen bubbles

Hydrogen bubbles

Anode

Water

Cathode

Power source

$$2H_2O \longrightarrow 2H_2 + O_2$$

Figure 33. Electrolysis
Forcing electricity through water can split water molecules into hydrogen and oxygen gases. The hydrogen stores energy.

The process is still used in South Africa today, where they convert coal and natural gas into diesel fuel in quantities sufficient for most of that country's needs.

Because the basic processes have been in use for decades, engineers have learned to improve efficiency and keep costs down. One recent study showed that gasoline could be produced from wind-generated electricity for the equivalent of $5.25 per gallon.[16]

Liquid fuels can also be made from high-carbohydrate plants like corn. Because the energy in corn comes from photosynthesis powered by the sun, the ethanol fuel that can be distilled from corn is considered solar-based. However, the fossil-fuel energy used to grow the corn using modern farming methods offsets most of the energy in the fuel, essentially making corn ethanol into another form of fossil fuel.

Corn-based ethanol also raises a significant moral issue. It converts human-edible food into fuel, when millions of people around the world are underfed. Already 40 percent of the US corn crop (which is 15 percent of global corn production) is diverted into ethanol production.

The increased demand for corn drives up food prices. A recent survey by the National Academy of Sciences estimated that globally biofuels expansion accounted for 20–40 percent of the food price increases that were seen in 2007–8, when prices of many food crops doubled.[17] The severe drought of 2012 reduced US corn crops and drove up corn prices,[18] generating calls for ethanol production to be cut back to relieve the market pressure.

Biofuels made from non-food sources avoid such problems. Several companies are trying to produce ethanol from switchgrass, a perennial grass that is inedible and grows on land unsuitable for other crops. The energy in switchgrass is primarily in the form of cellulose, which must first be broken down with enzymes into simple sugars for the distillation process. Other startup companies are converting cellulose from wood waste, and from corn-crop residue such as cobs, leaves, husks, and stalks.[19]

The National Research Council of Canada tested a jet fuel derived entirely from oilseed, a non-food crop that can grow in semi-arid regions such as the southern prairies in western Canada, where most food crops won't grow.[20] They found the biofuel to be a direct substitute for conventional jet fuel, and it produced less air pollution.

Research projects on advanced biofuels use genetically modified microbes to secrete fuel chemicals directly. Cyanobacteria have been genetically engineered to excrete chemical components of diesel fuel or ethanol when exposed to sunlight. The bacteria could be circulated in liquid solution through rooftop-mounted panels exposed to sunlight, negating the need for dedicated land areas.

Solar Energy Can Meet All Needs

The preceding sections describe many examples of how solar-based energy systems can meet modern energy needs. Can solar energy meet *all* of our energy needs? Human ingenuity has invented countless ways of applying energy to meet our purposes, so it might seem an impossible task to adapt all of those energy systems to use solar energy. But the number of forms of energy inputs to those energy systems is small: Electricity, heat, and fuels. Each of those energy forms can be produced from solar energy. The Principle of Energy Equivalence allows us to replace fossil fuels with such solar-derived energy.

That means there is no energy system that cannot be run from solar energy. Of course, the scale and cost of solar energy must be considered, and those I discuss in coming chapters, but there are no technical reasons why we could not run our lives and the entire world economy on solar energy.

In most cases, existing energy-using equipment could be retained, just substituting a new source of input energy. All electric motors run as well on solar electricity as they do on fossil-fuel electricity. Some energy applications will require adaptation. For example, cars that run on gasoline will need to be replaced with electric or plug-in hybrid electric cars so they can run on solar electricity instead of oil.

Converting from a fossil-fuel energy base to a solar energy base can be done gradually without disruption. When a business installs solar PV panels on its roof, it does not need to stop operating to make the transition. Most workers won't even notice that they have transitioned from a fossil-fueled enterprise to a solar-powered enterprise.

Replacing fossil-fuel energy with solar-based energy can be done on a case-by-case basis as opportunity and economics permit. That's because the energy source is already distributed. It pours down on the roofs of businesses and homes every sunny day. Deciding to make use of that currently wasted energy source can be done by individuals by installing solar equipment on the roof.

Each home and business enterprise can determine the best way to meet its energy needs with solar. Some will have enough exposed roof space or land area to generate all their own electricity. Others may generate more than they can use, selling the excess into the electrical grid for a profit. Those without adequate roof space can purchase solar electricity generated by others.

If the goal is to eliminate fossil fuel use for energy, and avoid a major commitment to dangerous nuclear energy, then we need to go 100 percent solar. Detailed studies from Brookhaven National Laboratory,[21] Worldwatch Institute,[22] University of California,[23] and Stanford University[24] confirm that all human energy needs can be met with solar-based energy sources.

It will take more than simply installing a lot of solar PV panels, though. The basic solar technologies like PV certainly work, as my own home demonstrates. But the characteristics of solar differ from fossil fuels and will require some adjustments to integrate it into our economies.

We are in a situation similar to when the first oil wells were drilled in the 1800s. An oil well can produce a lot of oil, but what do you do with that abundance? Raw oil was not very useful at first. The oil pioneers needed to learn how to refine it to extract useful components like kerosene, gasoline, and lubricants. Early transport of raw oil was primitive—they poured the oil into barrels and loaded the barrels onto horse-drawn wagons. A shortage of barrels became the bottleneck in the early days until tanker train cars and pipelines were developed to transport oil and its products. Also, oil didn't become a major energy source until a suitable heat engine was developed for it—the internal combustion engine. Only when all these components were in place could oil become a dominant energy source.

We are still in the early days of solar energy, and the systems for using it need to be completed. Certain components are already in place. Solar PV systems are now commonplace. Wind turbines produced 27 percent of the electricity used in the state of Iowa in 2013.[25]

But solar energy will need two further developments before it can fully take over from fossil fuels:

- Energy storage, including solar fuels.
- Energy efficiency.

The next two chapters describe what is needed for each of these components to enable us to shift to a 100 percent solar-based economy.

PART III: SOLAR IS THE ONE

CHAPTER 13

Energy Storage

There is a big challenge with solar energy: The raw solar energy source is variable by its nature. Solar energy is not available at night, and clouds can cut its availability at any time during the day. The solar resource also experiences seasonal variation, with more energy available in the summer with its long days and high sun angles than in winter with its short days and low sun angles.

Compare solar's variability to a coal-fired power plant that operates at the same rate for twenty-four hours per day. Critics of solar energy point out that reliable power is what our economy needs, so solar cannot meet all of our energy needs. They say we cannot run a global economy on an intermittent energy source.

That's like saying you can't take a shower unless it's raining outside. Our water-supply systems are designed to store water when it comes, and to make it available when it is needed. Water storage provides water that is available anytime on demand.

Energy-storage systems do the same for energy. They absorb energy when nature provides it, and give it back when we need to use it. The battery system on my solar home serves that role, making our power as convenient and reliable as that from the utility grid.

Energy storage is not new. The earliest users of fire gathered wood during the day for use at night. The wood pile was an energy-storage system, or "energy store." Here the term *store* does not refer to a place where you buy energy, but to an *energy stockpile* to be drawn upon as needed, like a squirrel's store of nuts.

When we graduated from the Wood Energy Epoch to the Fossil Fuel Epoch in the 1800s, we switched from wood energy stores to fossil-fuel energy stores. Since both systems were based on found energy stores, the transition from wood to fossil fuels was easy.

The transition from fossil fuels to solar energy will be more difficult. Fossil fuels differ from solar energy by having built-in energy storage. The original energy in fossil fuels came from sunlight that fell to Earth millions of years ago; this energy remains stored as chemical energy in the ground until we dig it up. The fossil-fuel energy store is a one-time gift of nature, which we've been drawing down since we discovered how to use fossil fuels.

To match the reliability of fossil fuels, solar energy needs *rechargeable* energy stores, like water reservoirs. Solar energy then becomes as reliable as water from a tap.

So putting up solar collectors is not enough. A complete system for using solar energy consists of collectors *and* energy storage.

If the storage part is left out, the transition to a 100 percent solar-powered economy is not possible. "Without technological breakthroughs in efficient, large-scale energy storage, it will be difficult to rely on intermittent renewables for much more than 20 to 30 percent of our electricity," said US Secretary of Energy Steven Chu in 2010.[1] With energy storage, fully 99.9 percent of our electricity needs could be met with solar-based energy sources by 2030, according to a University of Delaware study.[2] Recognizing this need, the California Public Utilities Commission established energy storage goals in 2013 for each electric utility it regulates.[3]

Because fossil fuels have built-in storage, research on energy storage has never been a priority. The power grid was not designed to accommodate a high level of renewables, but it should be, according to Haresh Kamath of the Electric Power Research Institute.[4]

The realization that energy storage is crucial for solar energy has stimulated development on many fronts. For example, the US Department of Energy established the Joint Center for Energy Storage Research (JCESR) in 2012 to coordinate research at the various national laboratories. Many private companies are also investing in research and development because they foresee a multibillion-dollar market for successful energy-storage systems.

The research is focusing on two areas: Storing solar electricity for the power grid, and making available portable stored energy for transportation. The following sections show how varied the approaches are to developing energy storage for these applications.

"Virtual" Energy Storage

Most solar PV installations on homes and businesses do not include any energy storage. Those systems rely on the utility grid as a virtual energy-storage system. When more electricity is generated than is needed, the extra is passed into the grid for someone else to use, and credits are granted. Those credits are used up at night or at other times when demand exceeds the solar supply. In such a net metering system, an end-of-year tally determines whether the system owner has a net utility bill to pay.

This arrangement shifts the burden of daily energy storage to the utility company, which can absorb a small amount of intermittent solar and wind electricity without creating instabilities in their system. Grid instabilities will arise when about 15 percent of the electricity in the grid comes from intermittent sources.[5] Germany is already experiencing such difficulties since about 27 percent of its electricity comes from renewable sources.[6]

Some utility companies have responded by increasing the interconnectedness of electric grids over wider geographic areas. They have found that one region's electrical peaks can fill in another region's electrical valleys. This method has proven most effective when mixing wind energy from areas with differing wind patterns.

Utility companies that have access to hydroelectricity can use that power source to even out the loads. Water stored behind a dam represents an energy store that can be turned on quickly to fill in the gaps from wind and solar sources. Because hydropower is another form of solar energy, this combination is particularly effective for increasing solar's share of the energy supply.

Utility companies can bring online other energy storage systems when they become available to better manage their loads, as described in the following sections. As more distributed PV and wind systems connect to the grid, utilities can provide their owners with centralized energy storage services, so that each site need not manage their own storage. Over time, the utility role may shift from electricity generator to energy distribution and storage hub.

Pumped Hydro and Compressed Air

Many utilities have taken hydropower a step further by turning it into a two-way system. A *pumped hydro* system consists of two lakes separated by a vertical height difference. Energy is stored by pumping water to the upper lake. Pumped hydro has a round-trip efficiency of about 75 percent, and can respond quickly to changing power loads.

Pumped hydro dates back to the 1890s, and is the largest form of grid energy storage working today. In the United States, about 2.5 percent of all electricity passes through pumped hydro, while Europe does 10 percent and Japan 15 percent.[7] But pumped hydro cannot be sited just anywhere, because it requires two lakes at significantly different elevations. Pumping water into an elevated storage tank is insufficient for utility-scale electricity.

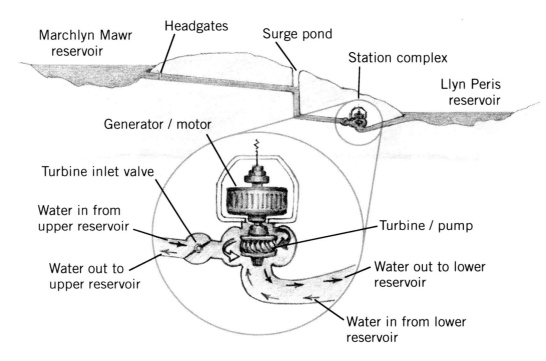

Figure 34. Dinorwig Power Station pumped hydroelectric storage system, Wales
Excess electricity powers an electric motor to pump water from the lower lake to the upper, converting the electricity to gravity energy for storage. To recover the energy, the process is reversed and water is run downhill through the turbine, changing the motor into a generator.

Using a system similar to pumped hydro, a small number of utilities use compressed air for energy storage. The system works by using excess electricity to compress air, storing the energy as elastic energy. When electricity is needed, the high-pressure air drives an air turbine to turn a generator.

But to store energy at a utility scale, high-pressure tanks are too small and expensive. So the compressed air is typically pumped into an underground cavern that is sealed to prevent air leaks. A new system being built by Pacific Gas & Electric in California will store compressed air in depleted natural-gas reservoirs.

A German utility company has operated a 290-megawatt compressed air energy-storage system for thirty years. There is also a 110 MW system in Alabama. This type of system can work only where suitable underground storage is available, which will limit its deployment.

Liquefying the air would eliminate the need for an underground cavern, making air storage feasible almost anywhere. At a pilot project built by Highview Power Storage in Slough, England, excess electricity runs a cryogenic refrigerator to cool air to -319°F, condensing the air into a liquid. The liquified air takes up far less volume and can be stored in insulated low-pressure tanks. To recover the energy, some of the liquid is warmed enough to regenerate pressurized air that can power a turbine. The system was invented by Peter Dearman, a garage inventor in Hertfordshire England, to power vehicles. It uses off-the-shelf components, and is expected to cost less than pumped hydro.[8]

Flywheels

Recent advances in material science have enabled *flywheels* to be reexamined for energy storage.

New flywheel designs use a carbon-fiber wheel spinning at up to 60,000 rpm. At those speeds, the edge of the disk travels faster than the speed of sound. To prevent burn up at such speeds, friction is eliminated by suspending the rotor on magnetic bearings and housing it in a vacuum. A magnetic coupling transfers energy in and out without touching the wheel.[9] Such flywheels can achieve 95 percent efficiency for energy storage (compared to 70–75% for batteries). Flywheels also do not deteriorate over time the way batteries do.

With a Department of Energy loan, Beacon Power built a utility-scale flywheel system in Stephentown, New York. It can store 18 gigajoules of energy and deliver it at the rate of 20 megawatts of power for fifteen minutes. In a move that bodes well for this technology, the private equity firm that bought Beacon Power plans to build a second such system in Pennsylvania.

Oddly enough, flywheel energy storage is showing up in Formula 1 racing cars. Recent rule changes permit Formula 1 cars to carry some kind of energy storage device to recycle braking energy produced going into a turn into acceleration coming out of the turn. Flywheels 30 centimeters in diameter are competing with batteries for the job. The winner

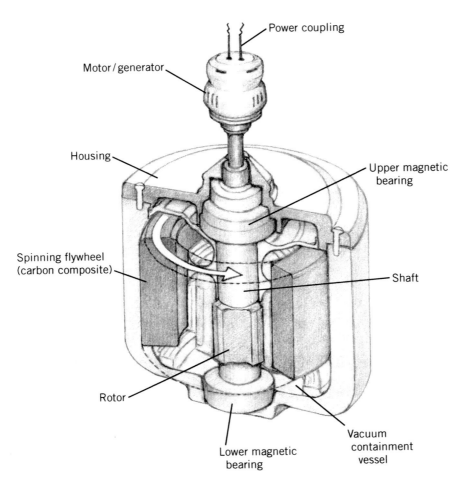

Power coupling

Motor/generator

Housing

Upper magnetic bearing

Spinning flywheel (carbon composite)

Shaft

Rotor

Lower magnetic bearing

Vacuum containment vessel

Figure 35. Flywheel energy storage

A flywheel stores energy as rotational kinetic energy in a fast-spinning rotor. When excess electricity is available, it spins up the rotor to high speed. To recover the energy, the rotor is connected to a generator to produce electricity.

of the 2012 Le Mans 24 race edged out his competitors using such a flywheel system.

Similar flywheels are being adapted to city buses, which constantly have to stop and restart, so recovering braking energy with a flywheel or battery can save fuel. London bus operator Go-Ahead has fitted Formula 1 flywheels to six of its buses to test the concept. Future hybrid cars may also replace their battery banks with flywheels, eliminating the need to periodically replace batteries.

Batteries

The most common way to store electricity today is with a battery. Strictly speaking, batteries don't actually store electricity. Rather, incoming electricity is converted to chemical energy at the battery's electrodes, and the battery stores that chemical energy. The battery can reconvert the chemical energy back to electricity on demand.

My home batteries store energy on lead plates immersed in liquid sulfuric acid. Almost every car on the road uses a lead-acid battery to start its engine. That technology is old, but relatively cheap and reliable. How old? Lead-acid batteries were used in the first electric cars around 1900.

Duke Energy's Notrees Battery Storage Project built for the 153-megawatt Notrees Wind Farm in West Texas uses giant lead-acid batteries because of their low cost and reliability. The battery can store 86 gigajoules of energy, which it can deliver at a maximum power of 36 megawatts. The battery system evens out the wind-energy supply, and makes the grid more resilient because it can respond almost instantly, faster than ramping up a fossil-fueled turbine.

Newer battery designs based on sodium-sulfur chemistry are even better suited for utility companies because of their high efficiency, long lifetime, and relatively low cost. For example, the sodium-sulfur battery system matched to the Rokkasho-Futamata Wind Farm in Hokkaido, Japan, can store 850 gigajoules, almost ten times larger than the Notrees lead-acid battery. But sodium-sulfur batteries are not for everyone. They contain highly corrosive molten sodium and sulfur kept at 570°F, and must be managed carefully.

Lead-acid batteries can be used in electric cars, but are not a good match for portable power because they are so heavy. My first electric car carried 1,100 pounds of lead-filled batteries, but I could only travel about thirty-five miles on a charge. New electric cars use lightweight lithium-ion batteries, a technology originally developed to reduce the weight of mobile telephones.

Lithium is one of the lightest elements, so batteries made from lithium are lighter per unit of energy stored.[10] A Tesla Model S electric car equipped with 1,200 pounds of lithium batteries can travel 265 miles on a charge, more than seven times farther than with a comparable weight of lead-acid batteries. An electric car with a range of 265 miles approaches the range of a gasoline-powered car.

Ultracapacitors

Quick recharge times are a specialty of *ultracapacitors*, which store static electricity directly on sets of thin metal plates that are separated from each other by nonconducting layers. Half the plates store positive charge and half store negative charge. Since there are no losses from converting to and from chemical energy, an ultracapacitor can be nearly 100 percent efficient.

Ultracapacitors can also be charged much faster than batteries. Because there is no chemical reaction, the charging rate depends only on how fast you can pump in electricity.

Ultracapacitors have longer lifetimes than batteries. A battery wears out because its chemicals eventually get polluted with extraneous chemical byproducts. Since an ultracapacitor uses physical storage of electrons instead of chemical storage, projected lifetimes are decades, not years.

Ultracapacitors show great potential, but current models have only a modest storage capacity. The basic problem is that pushing a lot of charged electrons into a small space creates enormous electric fields. As you continue to load electrons in, those electric fields get strong enough to break down the insulator, shorting out the device. Research is centered on finding insulating materials that don't break down in high electric fields.

Today's ultracapacitors can store only about 35 joules per gram of weight. Compare that to 120 joules per gram for lead-acid batteries, 900 joules per gram for lithium batteries, and 46,000 joules per gram for gasoline (which is *not* rechargeable). Ultracapacitors can be used today where you need fast and efficient storage of small amounts of energy. For example, some electric buses supplement their batteries with ultracapacitors for the regenerative braking cycle. They are more efficient than batteries, and do not wear out with frequent charging and discharging.

I own a cordless electric screwdriver that uses an ultracapacitor instead of a battery. It is one of the first consumer products to use an ultracapacitor, and it demonstrates the characteristics of all ultracapacitors. Unlike a battery, it reaches full charge in only ninety seconds, but unlike a battery, it will drive only about ten long screws before needing a recharge. It also tends to self-discharge faster than a battery. Fortunately, that quick recharge solves both problems.

Quick recharge is what electric cars need, so if future ultracapacitors can approach the energy-storage capacity of lithium batteries, they could replace the entire battery system in an electric vehicle. With an ultracapacitor, you could fill up your electric car as quickly as a conventional gasoline car today. That would make cross-country trips in an electric car possible once charging stations become standardized and common.

Flow Batteries

Until improved ultracapacitors become available, flow batteries could meet the same need. In a regular battery, energy is stored on solid plates immersed in liquid electrolyte, which facilitates current flow. In a flow battery, those roles are reversed—the energy is instead stored in the liquid electrolyte, and the plates just facilitate the current flow.

Since the energy is stored in the liquid part of a flow battery, charged liquid can be withdrawn and stored in a tank; this liquid is then replaced by fresh electrolyte from another tank. To recover the electricity, the flow is simply reversed.

By storing the electrolyte outside the battery case, you can increase the energy-storage capacity of a flow battery just by installing larger storage tanks for the electrolyte. Also, because the electrodes are not undergoing a chemical change, they don't wear out. Only the liquid electrolyte needs to be replaced as it degrades over time.

An electric car equipped with a flow battery could fill up as quickly as a gas-powered car. Instead of connecting an electrical cord to charge your electric car, you attach two hoses: One to pump out discharged electrolyte and one to pump in charged electrolyte. The car drives away in minutes with a full charge, while the charging station recharges the

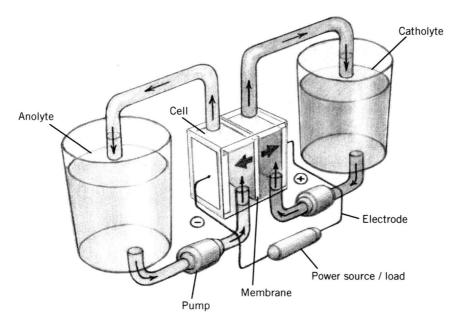

Figure 36. Flow battery

In a flow battery, energy is stored in the liquid electrolyte rather than in the battery plates. Once charged, the liquid can be withdrawn and stored in an external storage tank of any size. Large tanks can store large amounts of energy.

exhausted electrolyte at its own pace to make it available for the next customer.

Flow batteries are more complicated because they require pumps and sensors, but their operation can be almost entirely automated with computer control. Currently, utility companies are experimenting with flow batteries as they become commercially available. Two kinds of flow batteries are available today: zinc-bromine and vanadium redox. Neither has achieved great commercial success because the energy content of the charged liquid is low.

A new flow battery design from MIT provides higher energy content. In this *semi-solid flow cell*, a slurry holds tiny solid particles of battery material suspended in a viscous carrier liquid. The particles could be lithium that can be charged and discharged like a lithium battery. The solid particles enable it to carry ten times as much energy as other flow batteries.[11]

Hydrogen Fuel Cells

A fuel cell directly converts a fuel into electricity, so a fuel cell could also power an electric car. Of course, a heat engine connected to a generator can convert fuel into electricity, but a fuel cell accomplishes that feat without first burning the fuel for heat. Instead, a fuel cell manages a direct transfer of chemical energy to electrical energy.

When burning a fuel, the chemical reaction of combustion is not tightly managed, so the energy that's released ends up as random kinetic energy, or heat. Burning a fuel is trivially easy as it requires only ignition to proceed, but as it proceeds it randomizes the energy, making it less useful. In a fuel cell, the chemical reaction is carefully managed between a pair of electrodes, so instead of producing random heat, the chemical energy is converted directly to an electron flow in a circuit.

By not randomizing the energy, the maximum theoretical efficiency of a fuel cell is 83 percent, although existing fuel cells are only 40 to 60 percent efficient. That is still far higher than the 25% efficiency of a typical car engine.[12]

An electric car with a fuel cell would fill up on fuel like a gasoline car. Instead of burning the fuel, the fuel cell generates electricity to power the car's electric motor. A small chemical battery or ultracapacitor would be used to store energy from the regenerative braking cycle.

A fuel cell cannot run on gasoline, though. The chemical reaction at the electrodes can be managed only for a single chemical, and gasoline is a rich mixture of chemicals. Most fuel-cell development has used hydrogen gas as the fuel because it's a very simple chemical.

Hydrogen is a lightweight flammable gas that does not exist freely in nature. Hydrogen gas can be easily generated, though, by passing a large-enough electrical current through water. The electrical energy splits the water molecules, forming hydrogen gas on one electrode and oxygen gas on the other.

A hydrogen fuel cell essentially reverses that process, taking in hydrogen at one electrode and oxygen at the other, and generating electricity as it manages the chemical reactions. The electrical energy that went into making the hydrogen is regenerated by the fuel cell, and the waste product is just water.

A hydrogen fuel cell is like a battery connected to a fuel tank—it can produce electricity until the fuel supply is exhausted. An electric car powered by a fuel cell would replace long charging times with a quick fill-up of hydrogen fuel. If the hydrogen were generated from solar electricity, then the car becomes solar powered.

Hydrogen fuel cells have long been promoted for use in electric cars. The Hydrogen Fuel Initiative created by George W. Bush in 2003 poured over a billion US dollars into research on hydrogen-fueled cars for several years. But in 2009, the numerous unsolved technical problems led Energy Secretary Steven Chu to cut funding for the program.

The main problem with hydrogen is that it's a difficult material to store. Hydrogen exists as a gas at normal temperatures, so to store any significant quantity it must be compressed and stored under high pressure. Many people would be reluctant to drive around in a car containing a highly compressed canister of flammable gas. Hydrogen has long been associated with the Hindenberg, the hydrogen-filled German dirigible that burned spectacularly over New Jersey in 1937.

Hydrogen can be liquefied and stored at low pressure, but only at temperatures near absolute zero. The liquification process consumes 30–40 percent of the hydrogen's energy, and then special materials must be used to handle the extremely cold liquid (a rubber hose would freeze solid on contact). In storage, liquid hydrogen slowly boils off as heat leaks in, so the supply diminishes as it sits.

Hydrogen has an odd feature that makes it particularly difficult to handle: The hydrogen molecule is so small that it can actually dissolve into metals. As that takes place over time, the hydrogen interferes with the bonds between the metal atoms, making the metal brittle. Then you have a flammable gas in a high-pressure canister whose walls are becoming weaker over time—a formula for disaster.

The small size of the hydrogen molecule also makes it susceptible to leaking through the tiniest of gaps. Fittings that work with natural gas are insufficient for hydrogen. Leaks are a significant safety issue because hydrogen is colorless and odorless, and is explosive when mixed with air. Adding an odorant chemical as is done with natural gas to make leaks detectable by smell would not work with hydrogen because any odorant gas would leak more slowly than the hydrogen.

Utility companies are also reluctant to work with hydrogen as an energy storage medium. Utility companies have experience managing high-pressure natural gas pipelines, but those pipes were not subject to embrittlement from within. Also, the container volumes needed to store utility-scale energy as hydrogen gas are enormous, and power companies have little experience with large high-pressure tanks.

Liquid Fuel Cells

All of the problems that hold back hydrogen fuel cells are avoided with a fuel cell that runs on a liquid fuel. Instead of high-pressure gas tanks, an unpressurized liquid storage tank suffices. And a liquid fuel holds far more energy than the equivalent volume of pressurized hydrogen gas, in a form that does not easily leak or embrittle its container.

Prototype fuel cells have been developed for several liquid fuels, including butane, hydrazine (a component of rocket fuel), dimethyl ether (similar to propane), and methanol, which has emerged as the leading contender.

Methanol is a liquid alcohol similar to ethanol, but rather than being distilled from corn, methanol has traditionally been made from wood (hence its other name, "wood alcohol"), and more recently from coal or natural gas. Methanol is used today as an antifreeze, paint thinner, and racing fuel. Methanol has been used exclusively as the fuel in Indianapolis 500 race cars since 1965 because it is considered safer than gasoline in a crash.

For use with solar energy, methanol can also be manufactured from electricity, water, and carbon dioxide as described in the section titled "Solar Fuels" (page 150). If the electricity comes from solar sources, then methanol serves as a liquid solar fuel.

Methanol can be used in existing hydrogen fuel cells by adding a reformer, which takes in liquid methanol and chemically breaks it down to hydrogen gas. This approach takes advantage of all the research already done on hydrogen fuel cells, while avoiding the disadvantages of storing and handling volumes of hydrogen gas or ultra-cold liquid hydrogen.

Methanol can be more directly converted back to electricity by using an aptly named *direct methanol fuel cell* instead. Such a fuel cell operates directly from the liquid methanol, skipping the reforming step. Direct methanol fuel cells currently have low efficiencies and high costs, but have stimulated considerable research because of their significant potential.

A future electric car could pair a direct methanol fuel cell with a battery. The battery would power the electric motor to supply instant acceleration and handle regenerative braking, and the fuel cell would steadily keep the battery charged. Such an electric car could be "recharged" in minutes by refilling its tank with solar-produced methanol at a service station.

Georgetown University in Washington, D.C., already operates several transit buses on fuel cells powered by reformed methanol. They can refuel the buses in five minutes, and have measured near-zero emission levels. The buses carry a battery pack for acceleration and regenerative braking.[13]

A Methanol Economy

Liquid solar fuels such as methanol could also serve as the storage medium for utility electricity. When the supply of wind or solar electricity fed into a grid exceeds demand, utility company could convert the extra to liquid methanol. When demand exceeds supply, then fuel cells would convert the methanol back to electricity.[14]

Methanol is stable and does not self-discharge like a battery. Storage capacity could be scaled up by simply adding more or larger methanol tanks. With sufficiently large tanks, a utility company could cover entire cloudy days with solar-produced electricity.

Unlike pumped hydro or compressed-air energy storage, a liquid energy storage system does not require special geological formations, so it can be located almost anywhere. Since they operate quietly and without pollution, small substations could even be located in neighborhoods.

With very large storage tanks, the utility could create annual energy storage, saving summer energy for use in winter. Because winter days are shorter and the sun lower in the sky, less solar energy is available in winter, but winter is often when you need the most energy. Winter energy needs could be satisfied with summer energy, reducing the need for extra solar collectors in winter.

Methanol could also be used to transport solar energy from sunny climates to cloudy ones. Hauling methanol would be like hauling any other liquid petroleum product. Methanol tanker trucks could move solar energy from sunny Arizona to Seattle, for instance. Methanol tanker ships could transport solar energy from sunny equatorial countries to cloudy northern or southern countries. Methanol makes solar energy portable.

Nobel Prize-winning chemist George Olah uses the term *methanol economy* to describe a world where methanol is used for global energy storage and transport.[15] Methanol would serve as a common energy carrier, similar to the role electricity plays today. Methanol could be generated from a variety of sources and satisfy a variety of energy needs. Methanol would permit almost any new energy source to be incorporated, regardless of its nature.

For example, Carbon Recycling International built a commercial-scale methanol plant in 2011 near Grindavik, Iceland. The plant uses geothermal electricity to split water into hydrogen and oxygen, and then combines them with carbon dioxide to produce liquid methanol. The methanol is mixed with gasoline and sold in local filling stations.[16] In this way Iceland is replacing fossil-fuel imports with local geothermal energy for transportation. With further expansion, they expect to export methanol to other countries. The same model could substitute solar-generated electricity for geothermal energy and be used almost anywhere in the world.

CHAPTER 14

Energy Efficiency

When I set out to power my home with solar PV, I knew I had to work both ends of the problem. I had to position the PV panels in the best location to maximize my solar electricity, and I had to minimize my energy waste. Why spend money on panels only to throw away the electricity they produce?

To achieve the goal of a 100 percent solar-based energy system, we must get smarter about *using* energy. In the past we could throw unlimited quantities of cheap energy at a need. Such solutions are no longer appropriate because of rising energy prices and the environmental impacts of fossil-fuel energy systems. It's time to reexamine how we use energy, not just how much.

We can more easily meet all of our energy needs with solar by improving the overall efficiency with which we use energy. If we use less energy, then we will require less solar equipment to satisfy our needs.

For example, the United States consumed 3.9 trillion kilowatt-hours of electricity in 2010, which averages out to 444 gigawatts (444 billion joules per second) of power. Without energy-efficiency measures, we would need to average 444 gigawatts of solar-based electricity flow to be 100 percent solar. If the electricity demand could be reduced by half through efficiency improvements, then we would need to supply only

222 gigawatts of solar. We can reach the smaller goal sooner, thereby cutting off decades of extra carbon emissions. Since efficiency is almost always cheaper than new generating equipment, we'll save money too.

You can think of energy efficiency as another component of a complete solar-based energy system. Without energy efficiency, achieving a goal of 100 percent solar becomes unnecessarily difficult and expensive.

Let's be clear about what energy efficiency means. Energy efficiency does *not* mean doing without energy. You can save energy by turning off your home's furnace and sitting in a cold house, but that is energy denial, not energy efficiency. Energy denial means you forego some purpose in order to save energy.

Energy efficiency means accomplishing the same purpose but with less energy. Instead of *doing without*, it means *doing—without waste*. You still use energy to meet your needs, but you do it using a better energy system that doesn't waste energy, or at least wastes less. Instead of shutting off your furnace, you insulate and weatherstrip your home. You keep the same temperature and the same degree of comfort, but use less energy. That's energy efficiency.

Because energy can be measured, we can precisely determine when an energy system is being efficient and when it's not, whether it's a heater, a car, or a factory.

The Principle of Conservation of Energy stipulates that energy cannot be destroyed, so it can always be accounted for in some form. For example, input to a car's energy system is in the form of gasoline chemical energy, and the output is kinetic energy of the moving car and heat energy. If you can count all the joules of input energy and all the joules of output energy, then the amounts will exactly match if you're careful. The ledger will always balance when accounting for energy.

But the heat energy is not useful to your purpose of moving about. Energy accounting lets us classify the transformed bits of energy into two columns of figures: One column for useful energy and another for useless energy. The kinetic energy of the car goes into the useful column, because that was our intended use for the gasoline. The heat generated by tire and air friction goes into the useless column, as does the heat pouring out the exhaust pipe, because none of those contributes to your purpose for using the gasoline.

Once you've sorted the transformed energy into the two columns, the efficiency can be precisely determined. Efficiency is defined as the percentage of the input energy that ends up in the useful column. In other words, what percentage of the energy you buy ends up actually serving your energy needs?

In the case of a car's energy system, not very much.[1]

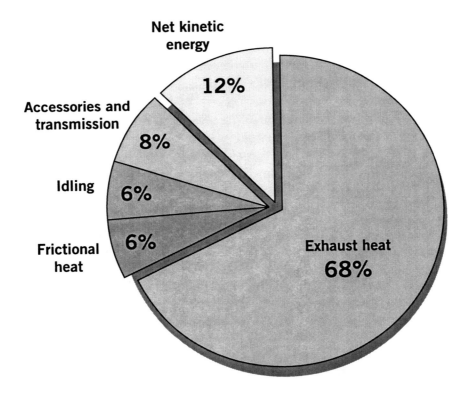

Figure 37. Net efficiency of a fossil-fueled car

Only 10–12 percent of the energy in automobile fuel results in motion. Fully two-thirds of the energy is wasted as heat flowing out the exhaust pipe.

- Best practical efficiency of the internal combustion cycle is 32 percent (heat engine limitation), leaving 68% of the energy as heat flowing out the exhaust pipe.
- Frictional losses lower it further to 26 percent.
- Idling in urban traffic lowers it further to 19–20 percent.

- Accessories and automatic transmission lower it further to 10–12 percent.

For every ten joules of gasoline energy you buy, only about one joule ends up in the useful column. So a gasoline-powered car would be rated as only 10–12 percent efficient as a transportation system. With such low efficiency, cars have great potential for improvements in energy efficiency.

The biggest loss comes from using a heat engine to power the car. Low efficiency is an unavoidable characteristic of a heat engine because it randomizes the energy as heat before trying to extract the useful part. The first steam engines were only about 1 percent efficient. In its primary application of pumping water from a coal mine, that low efficiency was acceptable for two reasons: plenty of cheap coal for the steam engine's boiler was available from the mine, and it was cheaper to operate than the teams of horses that it replaced.

Electric vs. Gas-Powered Automobile

An efficient gasoline-powered car might get 35 miles per gallon. At $3.50 per gallon (2014 US price), that amounts to 10 cents per mile in energy cost. An efficient electric car might get 4 miles per kilowatt-hour of electricity. At $.12 per kilowatt-hour, that is only 3 cents per mile in energy cost. When compared to the 10 cents per mile for gasoline, driving an electric car is a bargain.

Heat engine efficiency has certainly improved in the last 200 years, but even modern steam turbines used for utility power generation are only about 35 percent efficient. The best engineering practices cannot overcome the fundamental limitation of heat engines. Engineers are bumping up against the theoretical limit of converting heat to kinetic energy, so there's little room for improvement. Once you randomize energy by converting it to heat, it's very difficult to get it all back to highly organized mechanical energy.

An electric motor, which does not operate on heat, is about 90 percent efficient in converting the incoming electrical energy to kinetic energy. The high efficiency of electric motors makes an electric car cheaper to operate than a gasoline car, even though electrical energy costs about 30% more than gasoline per unit of energy.

Most people do not consider themselves profligate wasters of energy. The general public knows that wasting energy wastes money and is generally bad for the planet. Yet the energy systems we all use are not particularly efficient, so just by going about our everyday business with our current energy systems, we waste energy. Most cars use heat engines with a net efficiency of 10–12 percent or less. Most electricity is generated using heat engines that are 35 percent efficient or less. By simply participating in modern society, you use its inefficient energy systems.

Existing energy-using devices have low efficiency because they were developed during a time of cheap fossil fuels. Why spend money to make a system more efficient when energy is so cheap? Now that energy is getting more expensive, it makes sense to spend a bit more up-front to save energy in the long run. The traditional incandescent light bulb is a good example.

The electrical current flowing through a light bulb's filament heats the filament to thousands of degrees, which causes the filament to glow with light. But using heat to make light is quite inefficient because most of the energy ends up as heat or infrared radiation, which is not visible and does not contribute to the bulb's purpose of lighting a room. Most incandescent bulbs are less than 5 percent efficient at converting electricity to visible light. They are so inefficient that many countries are requiring them to be phased out over time.

A compact fluorescent light uses one-fourth the energy of an incandescent bulb for the same light output. Both bulbs serve the same purpose of lighting a room, but one uses far less energy to accomplish the task. That's a classic example of how replacing an older energy system with a more efficient one can reduce the amount of energy needed for that purpose. If your electricity comes from fossil fuels, that means one-fourth the carbon dioxide will be emitted. For solar-generated electricity, it means reducing the size and cost of the solar panels.

Compact fluorescents cost more initially, but they save money in the long run by using one-fourth the energy and lasting eight to thirteen times as long as conventional bulbs. LED lights are even more efficient and long lasting than compact fluorescents.

Studies by the Rocky Mountain Institute and others have shown that modern society has great potential for improving the efficiency of our energy systems, from lighting to heating to transportation to industrial processes. These studies show that a full installation of efficient electric-

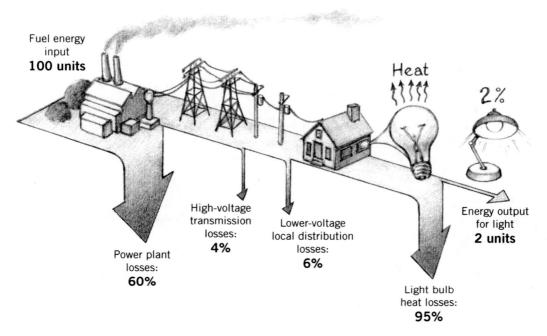

Fuel energy
input
100 units

Heat

2%

Power plant
losses:
60%

High-voltage
transmission
losses:
4%

Lower-voltage
local distribution
losses:
6%

Light bulb
heat losses:
95%

Energy output
for light
2 units

Figure 38. Chain of energy conversions
Saving energy at the point of end-use saves energy in all the preceding stages. One unit saved at the end can save ten units for the whole chain.

ity-using devices across the board could save 75 percent of US electricity.[2]

Improvements in energy efficiency by the end user are the most effective. That's because most energy consist of a chain of energy conversions, each with its own losses.

Also, end users know the most about their own energy use. Individual people, businesses, and institutions know best how energy is applied to meet their needs, and how upgraded equipment that's more efficient might fit into those needs. Improving the energy efficiency of our society will involve millions of individual decisions.

Some of the biggest energy wasters in modern homes today are *phantom loads*, devices that consume electricity even when they're supposedly turned off. Printers, fax machines, and DVRs are typically left on in order to provide on-demand service. Most drop into an energy-saving low-power mode when not in actual use, but still consume a small amount of electricity in that mode in order to stay "awake."

Any device with a remote control, such as a TV or stereo, has to remain in standby mode to receive the signal from the remote. In standby mode, the device may consume only a few watts, but it does so twenty-four hours per day. Also, any device with a clock, such as a stove or microwave oven, also remains partially on all the time to power the clock.

Other phantom loads come from cordless electronic devices such as cell phones, music players, and cordless tools such as drills. A charging cube or station continues to draw a small amount of power when the device is fully charged, and even when it's not plugged into the charger at all. If a charger feels warm, then it's using electricity.

Each of these devices may use only a small amount of electricity, but together they add up. A 2008 study found that the average California home contained more than forty products constantly drawing power. Together, those products consumed nearly 1,000 kilowatt-hours per year in their standby mode, representing 8 percent of the average household electricity use.[3]

In our solar-powered home, we have eliminated most of these phantom loads by putting such devices on switched power strips. When the power strip is switched off, the devices connected to it are truly off, and consume none of our precious solar electricity. For battery chargers, we use a timer to switch off the charger when the batteries are full (our timers have zero phantom load).

Home heating and cooling systems have huge potential for greater energy efficiency. A typical old gas furnace installed when energy was cheap might be 50 percent efficient, while new gas furnaces are up to 98 percent efficient. New low-budget homes employing only cost-effective efficiency technologies can be built to use half the energy compared to conventional homes. Including passive solar features in the home design can reduce heating demand by 80–90 percent.[4]

Super-efficient homes can be built with extra-thick insulated walls and careful attention to sealing all cracks to reduce outside air infiltration. Such homes can be mostly heated by the heat given off by the people, lights, and appliances in the building. Fresh air is provided by an Energy Recovery Ventilator, a device that heats incoming fresh air with outgoing stale air without mixing them. Such a home can be built more cheaply than a conventionally heated home by completely eliminating costly centralized heating and air-conditioning systems.[5].

If you combine an energy-efficient building with a solar PV array on the roof, you can achieve a *net-zero-energy building*. The basic idea is that the PV array annually generates as much energy as the building uses, resulting in no net use of fossil fuels. The net-zero goal can generally be reached only if the building is highly efficient in its energy use. Such a building models on a small scale how our society can move to a 100 percent solar-based economy by combining efficiency measures with solar-based energy.

In the United States, Presidential Executive Order 13514 from 2009 mandates newly constructed federal buildings to achieve net-zero energy by 2030. The American Institute of Architects and the state of California have also called for new commercial buildings to be net-zero energy by 2030.[6]

Most commercial buildings operate below their optimum level of energy efficiency. A study by Lawrence Berkeley National Laboratory showed that simply "tuning up" the existing energy systems in commercial buildings could save an average of 16 percent of their energy use, reaching up to a maximum of 30 percent.[7]

Most industries also have great potential for improving energy efficiency. With the history of low fossil-fuel prices, industry has typically treated energy costs as a minor expense relative to other costs of doing business. Those business that actually pay attention to energy efficiency have saved 30 percent or more. Here is a story from Rocky Mountain Institute that industrial shareholders should pay attention to.

In 1981, Dow Chemical's 2,400-worker Louisiana division started prospecting for overlooked savings. Engineer Ken Nelson set up a shop-floor-level contest for energy-saving ideas. Proposals had to offer at least 50 percent annual return on investment. The first year's twenty-seven projects averaged 173 percent return on investment. Startled at this unexpected bounty, though expecting it to peter out quickly, Nelson persevered. The next year, thirty-two projects averaged 340 percent return on investment. Twelve years and almost 900 implemented projects later, the efforts had averaged (in the 575 projects subjected to the *ex post* audit) 202 percent predicted and 204 percent audited return on investment. By 1993, the whole suite of projects was paying Dow's shareholders $110 million every year. [8]

As the Dow story shows, most energy-efficiency measures are highly cost effective in the long run, yet most have not yet been carried out. Why? Getting people to pay attention to energy seems to be the biggest hurdle. According to Amory Lovins,

> Increasing energy end-use efficiency—technologically providing more desired service per unit of delivered energy consumed—is generally the largest, least expensive, most benign, most quickly deployable, least visible, least understood, and most neglected way to provide energy services.[9]

The reasons energy efficiency is so neglected include:[10]

- Not knowing—most people don't know which energy efficiency measures can pay for themselves in a few years.

- Not caring—energy costs are small enough to tolerate some inefficiencies.

- Not owned—landlords who don't pay utility bills have little or no incentive to install efficiency improvements. Approximately one-third of residential buildings are occupied by renters.

- Not motivated—doing nothing is easier than doing something.

Many people are also demotivated to save energy when they see so much energy being wasted around them. Let's say you install all LED lights in your home, and carefully turn off lights when you leave a room so as to not waste electricity. Then you drive by a shopping center after hours and notice that most of their lights are on, and they remain on all night. You might wonder "Why should I bother saving my small amount of electricity when someone else is just going to waste it?"

This situation changes remarkably when you start to make your own energy, such as when you install your own solar PV system as I did. I paid a lot of money to produce my electricity, and I will not tolerate wasting it. The electricity I generate feels far more precious than utility electricity, so I do all I can to maximize its use.

Improving efficiency takes on greater urgency if the goal is to replace 100 percent of our fossil-fuel systems with solar-based systems. Efficiency can get us halfway to a solar-powered economy by letting us work both ends of the energy problem. We bring down the size of the energy flows through efficiency improvements, and we bring up the solar electric systems to cover the rest. Somewhere in the middle those two efforts will meet, at which point we will have arrived at a 100 percent solar-powered economy.

Efficiency also buys us time. The price of solar energy systems continues to drop, and will eventually be cheaper than all forms of fossil fuels. Until then, efficiency measures can halt the growth of fossil fuels, allowing solar to catch up. The transition to solar will be realized sooner if solar isn't chasing a moving target.

With greater efficiency, we avoid making the wrong investments in our energy systems. Utility companies facing growing energy demand will seek the conventional solution: Build more fossil-fueled power plants. Such power plants have an expected lifetime of at least thirty years, so each new plant is a commitment to burning fossil fuels for decades. If efficiency can ease the growth rate, then those fossil-fuel investments can be avoided, thereby avoiding thirty years of their associated carbon dioxide emissions.

If you need confirmation that we can live well on less energy, just compare the energy use in America to Europe and Japan. Energy use per person in Germany and Japan is about half that of Americans, for a similar lifestyle. For example, in 2011, annual energy use in Germany averaged 156 gigajoules per person, while it was 296 gigajoules in America.[11]

Some economists have claimed that the rate of energy use tracks a country's gross domestic product (GDP), implying that greater energy use leads to greater prosperity, but GDP is a poor measure of the quality of life. GDP measures all kinds of economic activity, good and bad, and all activities require energy.

Better quality-of-life indicators include infant mortality rate, life expectancy, educational opportunities, community engagement, leisure time, and political freedom. These can be combined into alternative measures such as the Human Development Index[12] or the Genuine Progress Indicator.[13] More careful energy studies show that only infant mortality and life expectancy are dependent on energy, and only in developing countries. Higher energy use beyond basic modern improvements does not bring greater cultural flowering, does not promote social stability, and is not even the precondition for greater economic prosperity.[14]

CHAPTER 15

Solar PV Will Be the Biggest Solar

Energy storage and energy efficiency will make it possible for solar-based energy sources to take on the entrenched fossil fuel energy industry. Energy storage will remove the barrier of solar's intermittency, and energy efficiency will bring down the demand for energy to a more reasonable level.

To actually supply the remaining energy flows from solar sources, we'll need to install a lot of solar equipment over the coming decades. But which equipment? Solar energy comes to us in many forms: Direct sunlight, wind energy, wave energy, elevated water, and photosynthesis.[1] Each form requires specific equipment to divert the natural solar energy flow to a form suitable for human purposes. For example, to put wind energy to use, we install a wind turbine and generate electricity. From direct sunlight we can use PV to produce electricity, flat panels for water heating, or concentrating solar mirrors for industrial heat.

Given the diverse forms of solar-derived energy and the wide-ranging ingenuity of humans, the future will likely see a mix of solar-based technologies supplying power to human civilization, but it's unlikely that all solar forms will be exploited equally. Most will run into limitations as they're scaled up.

One important lesson learned from the history of fossil fuels is that what works on a small scale can create problems on a large scale. When cars were few in number, the pollutants and carbon dioxide in their exhaust were diluted to insignificant levels in the atmosphere. But when millions of vehicles hit the road, their collective exhaust overwhelmed the atmosphere, causing smog on a local level and climate change on a global level.

What happens when solar technologies are scaled up to meet the huge global energy requirements of modern society? Can any of the seemingly benign solar energy sources scale up to global levels without creating its own raft of problems?

Of the solar technologies available today, only PV can scale up sufficiently to enable us to reach 100 percent solar. A 2015 MIT study concluded that solar PV is "one of the few renewable, low-carbon resources with both the scalability and the technological maturity to meet ever-growing global demand for electricity."[2]

Hydropower has been producing clean electricity for over a century, but hydropower has its own environmental problems. For example, the multiple dams on the Columbia River in the American Northwest block salmon migration so completely that wildlife managers have resorted to transporting migrating fish by tanker boat to keep the fish population from crashing.

Many people also object to the permanent flooding of large tracts of farmland or wildlife habitat behind a dam. The reservoirs behind the world's large dams (higher than fifteen meters) cover about 500,000 sq km, about the area of Spain.[3]

Only certain sites meet the criteria for building a dam, further limiting its potential for growth. Europe has already exploited 63 percent of its technically feasible hydropower capacity, and North America 72 percent. Significant potential for hydropower dams still exists in Asia, Africa, and South America.[4] Hydropower will grow there if it can overcome the significant ecological and cultural barriers to altering river flows.

Dam-free hydropower using turbines immersed in a running river can be fitted to many more sites, but they are largely unproven, with unknown long-term environmental impacts and comparatively small energy outputs.

Wind power works today and will likely be a significant contributor to future energy needs, but it is also limited by its siting requirements. A wind farm must have reliably high winds, so only specialized sites will do. Objections based on noise or visual impacts have slowed or halted some projects, forcing them away from population centers. Offshore wind farms avoid such problems and are being developed by Britain, Denmark, Germany, Holland, and other countries.

Wave power remains experimental and is limited to coastal sites. Its environmental impacts have yet to be determined because no single design has emerged as a clear leader.

Biofuels suffer from the basic inefficiency of photosynthesis, which converts only about 1 percent of the sunlight into usable chemical energy. Thus large land areas must be dedicated to biofuel crops if they are to significantly displace fossil fuels. Humans already consume nearly 30 percent of all the Earth's photosynthetic energy production.[5] Massive production of biofuels would inevitably compete with food crops, forcing a moral choice between energy use and food sustenance.

PV panels have the advantage of not requiring dedicated land area. Since PV panels have no moving parts and make no noise, they can be installed on rooftops or on overhead racks above paved areas. PV fits into urban settings like no other energy source. PV is completely modular, so a small installation can be as efficient as a large one, which means PV can be put just about anywhere that has clear exposure to the sun. And direct solar energy is much more evenly distributed over the planet than wind, wave, and hydro energies, making PV usable in almost all areas inhabited by humans.

For these reasons PV has the greatest potential to grow large enough to displace fossil fuels. Other solar technologies will no doubt contribute, but the dominant provider of energy will likely be PV. Before committing to such a major shift, however, it is worth asking what are the positives and negatives of embarking on this path. The next sections go into greater detail on why solar PV is the best choice for our energy future, and describe what problems might arise if we commit to it.

The Many Positives of Solar PV

Solar PV has at least twenty positive aspects that make it a compelling choice for our future energy system. Keep in mind that while other solar-based energy sources share some of these positive features, only solar PV has them all.

1. Solar PV does not contribute to global warming.
2. Solar PV does not acidify our oceans.
3. Solar PV does not pollute our air.
4. Solar PV does not pollute our water.
5. Solar PV is safe.
6. Solar PV does not produce radioactive waste.
7. Solar PV is secure energy.
8. Solar PV saves oil for use as a raw material.
9. Solar PV saves water for other uses.
10. Solar PV is simple.
11. Solar PV is available to all.
12. Solar PV is distributed.
13. Solar PV is reliable.
14. Solar PV will not run out.
15. Solar PV is a solid investment for business.
16. Solar PV is good for local economies.
17. Solar PV is good for national economies.
18. Solar PV scales well.
19. Solar PV requires no dedicated land.
20. Solar PV is ready now.

Each of these advantages is explained more fully in the following sections.

1. Solar PV Does Not Contribute to Global Warming

Solar PV systems emit zero carbon dioxide during operation. Thus solar PV can shut down the primary driving force behind global warming and climate change.

What about the energy used to make PV panels? If fossil fuels are used to make PV panels, then those fossil fuels will emit carbon dioxide. Studies have shown that it takes two to two and a half years of operation for a crystalline silicon PV panel's output to match the energy used to make the panel.[6] With an estimated lifetime of thirty years, that means more than 90 percent of the energy produced by a PV panel is entirely free of pollution. If solar panels are instead manufactured using solar electricity, in a kind of PV "breeder" factory powered by PV, then the PV panels coming out of that factory are 100 percent carbon-free.

2.   Solar PV Does Not Acidify Our Oceans

By eliminating carbon emissions, solar PV systems also do not contribute to ocean acidification. Since acidification is a direct result of human carbon emissions, and since the emerging consequences of acidification are so drastic, this benefit of solar PV carries particular strength.

3.   Solar PV Does Not Pollute Our Air

Solar PV systems emit zero air pollution during operation. Because they consume only pure light and no chemical fuel, they emit no hydrocarbons, ozone, particulates, carbon monoxide, nitrogen oxides, and sulfur oxides as fossil fuels do. Those pollutants harm human health by increasing suffering from chronic bronchitis and asthma, and increase the risk of cardiovascular problems. Solar PV would also eliminate the airborne mercury emitted by coal power plants, as well as acid rain.

Visible smog would also be greatly reduced since most of it comes from burning fossil fuels. Battery-electric vehicles recharged with solar PV would largely eliminate air pollution from the transportation sector.[7] When all the cars and trucks in major cities are solar-charged electrics, the skies over those cities will clear and their residents will breathe more easily.

Of course, the manufacturing processes for solar PV need to be examined for air pollution as well. Processing photovoltaic materials requires using many volatile organic chemicals, but PV plants already control such pollutants under existing clean-air regulations. At the end of the panels' lifetime, the panel material can be recycled.[8]

4. Solar PV Does Not Pollute Our Water

Unlike fossil fuels, solar PV does not contribute to water pollution. PV panels are sealed in glass, so rainwater never comes in contact with the active material. The only potential for water pollution can come during manufacturing of the panels.

The semiconductor industry has proactively set its own water-quality standards through the SEMI® industry association,[9] with member companies in North America, Europe, and Asia. Companies that adhere to the SEMI-PV3-0310 standard[10] treat any water that is to be released. New companies that have not yet invested in onsite treatment ship the waste to a treatment facility.

By displacing fossil fuels, solar electricity will eliminate the water pollution routinely generated by oil, coal, and natural gas production, whose devastating effects are described in the section titled "Ten Reasons Why Fossil Fuels Are Not Good For Us Anymore" (page 81).

5. Solar PV Is Safe

Solar PV has another important benefit—it's completely safe. Under normal operation, a properly installed PV system poses no hazards to humans and other living things. Certainly solar electricity can be dangerous if the system is damaged, so firefighters are updating their procedures to handle potentially live PV panels and high voltages.[11]

In contrast, the routine operation of every nuclear power plant requires constant attention to safety protocols. Billions of dollars are spent on safety features, with frequent inspections to ensure their operation. These costs are necessary to protect the public from exposure to radiation. Yet despite these expensive measures there are still leaks and accidents.

A solar deployment would save lives in the coal industry as well. Coal mining has historically been one of the most hazardous industries to work in, with over 100,000 miners killed in the United States alone during the 20th century.[12] Modern safety methods have reduced that toll in the U.S. to an average of thirty deaths per year, but such methods are not universally practiced worldwide.

6. Solar PV Does Not Produce Radioactive Waste

The biggest safety issue of nuclear power has yet to be faced. The long-term storage of highly radioactive waste from the power plants has not even begun. The scientific and political problems of siting a nuclear waste dump have not been solved despite forty years' effort and expense. And once a dump accepts waste, it must be monitored for thousands of years to prevent exposure to our descendants.

If nuclear energy replaces fossil fuels on a large scale, then there would have to be hundreds of truck trips per year transporting high-level radioactive waste through populated areas, with a risk of accidental release of radioactive material. With large quantities of such material in circulation, it would be only a matter of time before some of it was stolen by terrorists and used in a dirty bomb.

Solar electricity would not eliminate the need for a long-term nuclear-waste storage site, because we have already generated tons of the stuff. But solar electricity would halt the growth of that stockpile, and reduce the exposure of the population to radioactive transports and nuclear terrorists.

7. Solar PV Is Secure Energy

Because all economic activity is powered by energy, all nations are concerned about their supply of energy. Energy security colors national policies and international relations far more than any other commodity. Oil is the one commodity that no nation can do without.

> A quick examination of the world history over the last century would reveal the fundamental impact access to crude oil has had on the geopolitical landscape. Fortunes are made and lost over it; wars have been fought over it.
>
> —US National Research Council[13]

Oil divides the world's nations into "haves" and "have-nots." Those nations with oil wield considerable power over those that must import oil to keep their economies going. Imports are subject to cutoff for political or economic reasons, as was amply illustrated by the 1973 world oil embargo.

Importing nations distort their principles and policies to maintain friendly relations with the oil-exporting nations that they are dependent upon. Why else would the United States, one of the world's oldest democracies, diligently maintain kings in power in Saudi Arabia, and deploy its army to restore the Kuwaiti royal family to power in the 1991 Gulf War?

A transportation system based on solar-charged electric vehicles breaks that dependency on oil. Because the sun bathes every country in solar energy, all countries become solar "haves" that can rely on a secure supply of energy within their borders.

The other side of energy security is protecting energy infrastructure such as power plants, transmission lines, and refineries from natural disasters and terrorist attacks. A single attack that disables a large centralized power plant can put millions of people in the dark. The wide dispersal of distributed solar PV systems makes them harder to target.

8. Solar PV Saves Oil for Use as a Raw Material

Oil is usually viewed as an energy source, but oil also serves as a raw material for many chemical products, including pharmaceuticals, solvents, fertilizers, pesticides, and plastics. About 16 percent of oil production goes to materials rather than energy.[14]

If we burn all the oil for energy, then future generations will not have this valuable material resource. How much we preserve will depend on how quickly we can transition from oil to solar-based energy systems.

9. Solar PV Saves Water for Other Uses

Most people are surprised to learn how much water our current energy systems use. In coal and nuclear power plants, water is used for cooling. For a heat engine to continue to operate, its waste heat must be removed.

Some power plants are sited so they can divert river or sea water through the plant to carry away the heat. In the United States, 41 percent of all the freshwater withdrawn from lakes and rivers goes to this energy cooling.[15] Such pass-through cooling warms the water but does not consume it. However, the power plant is dependent on the water being available, and may have to cut back during periods of low water flow or high water temperatures. During the 2003 heat wave in Europe, seventeen nuclear reactors in France had to be curtailed or shut down to prevent overheating.[16]

Most power plants that are not able to use river or sea water instead use evaporative cooling in those huge conical towers you often see near a power plant. Heated water sprayed into a fine mist inside the tower evaporates and carries the waste heat into the atmosphere. This process consumes water—more than half a gallon of water for each kilowatt-hour generated.[17] Large power plants consume billions of gallons of water each year.

The unconventional drilling methods that are propping up the oil and gas supply also consume enormous amounts of water. Chesapeake Energy Corp. reports that drilling a deep shale gas well requires between 65,000 and 600,000 gallons, but that pales compared to the water required for the fracking process to break up the rock—an additional 4.5 million gallons on average.[18] In the United States, 30,000 shale gas wells were drilled during 2013.[19]

By contrast, a solar PV system uses no water for its routine operation. Large PV farms will use water to wash the collector surfaces to maximize output, but on average that process uses only 0.026 gallons per kilowatt-hour.[20] Wind farms have similar low water needs, but large solar thermal plants that use mirrors to drive a heat engine may consume water at the same rate as a conventional power plant if they use evaporative cooling.

With water stress growing in many parts of the world, energy systems like solar PV that do not compete for water gain an advantage over those that do.

10. Solar PV Is Simple

Although manufacturing solar PV panels employs complex modern technology, generating electricity with the panels is elegantly simple. You just mount a panel in a location exposed to regular sunshine, connect the wires to an inverter, and you have a useful source of electricity.

Solar PV's simplicity can make installing it a do-it-yourself project. For example, the Plymouth Area Renewable Energy Initiative (PAREI) in New Hampshire organizes solar "barn raisings," where neighbors and installers get together to install projects themselves.[21]

Solar's simplicity also permits easier entry into the business. Becoming a professional solar PV installer requires a few months of on-the-job training, in contrast to the multiyear university degree required to become a nuclear engineer.

11. Solar PV Is Available to All

The sun conveniently spreads solar energy over the face of the Earth. That means all countries have access to solar energy, though with some variance in the amount available in each country due to latitude and climate. Solar is an energy source for all humankind.

Because energy is the basis of all economic activity, solar PV enables more people to engage in economic activity. Residents of countries where electricity service is sporadic have difficulty keeping a business going. If everyone has energy, then everyone can participate in the economy.

Solar PV can also bring modern energy to the billion or so people in the world who don't currently have access to electricity. Many developing countries cannot afford to extend transmission lines from centralized power plants to every remote village. A small PV system with battery energy storage can provide the first electricity service for such villages. Small amounts of electricity for lights and refrigeration can make a huge difference in the lives of village residents.

Evans Wodongo from Kenya recognized that opportunity. He designed a solar-powered lantern he calls MwangaBora! (swahili for "Better Light") and distributes them to poor villages. Each lantern has a battery, LED light, and PV cells for recharging.[22] These simple, inexpensive lights are having all these effects:

- The lantern replaces the use of expensive kerosene for lighting, enabling people to spend more money on food.
- It enables the use of light for longer periods at night so children can study, thereby improving education.
- It eliminates indoor air pollution from burning kerosene, reducing respiratory illness induced by kerosene's soot particles.
- It eliminates the fire hazard of kerosene in wood and thatch huts.

Because solar PV is modular, these village energy systems are growing incrementally. Some will reach full electric service without ever connecting to the fossil-fuel-powered grid. They can skip the fossil-fuel age completely and jump directly to sustainable energy.

12. Solar PV Is Distributed

Distributed energy is defined as electricity produced from many small sources distributed over a wide geographical area. Solar PV is distributed, while nuclear and coal-fired power plants are not, because their economics require them to be large and centralized.

Distributed energy systems have these advantages:

- Shorter distance to end user so less energy is lost in transmission lines.
- Not susceptible to grid-wide failure due to natural disaster or terrorist attack because there is no single centralized facility.
- No need for major new transmission lines and their dedicated corridors of land, which are becoming harder to secure.
- Can use smaller distributed units for energy storage, which can be standardized and mass produced to make them cheaper.

The distributed nature of solar PV will require a *smart grid*, which is already being developed. A smart grid differs from a regular electricity grid because *information* flows over the lines in addition to power. The information can be used by grid managers to match myriad smaller power sources to loads in the most efficient manner, routing energy to and from distributed energy storage systems as needed. Worldwide investments in smart-grid deployment totaled $13.9 billion in 2012.[23]

13. Solar PV Is Reliable

At first glance, solar PV would seem to lack reliability, since a passing cloud can cut off the source of energy. Just as my batteries enable my home PV system to operate twenty-four hours per day, seven days per week, so will utility-scale energy storage enable the electricity grid to operate 24/7. Energy storage makes the intermittent nature of solar energy irrelevant.

Adding energy storage makes solar-based energy more reliable than fossil fuels, whose reliability depends on maintaining a continuous supply, an increasingly risky proposition. The reliability of solar-based energy will depend on adequate energy storage. With energy storage, solar can provide electricity for base loads, intermediate loads, and peaking loads.

PV panels themselves have already proven to be highly reliable. They are solid-state devices with no moving parts. With no moving parts, there are no worn parts, no lubrication of parts is needed, and no replacement parts are required. They don't wear out from the sun's ultraviolet light as most plastics do. PV panels are so reliable that PV manufacturers typically guarantee them for twenty to thirty years. Such long guarantees are unheard of for other consumer products. My own PV panels are over fifteen years old and show no signs of degradation.

14. Solar PV Will Not Run Out

Solar energy has another kind of reliability that fossil fuels can never match, and that is the long-term viability of the energy source. Fossil fuels exist on our planet in a finite supply, which we are gradually consuming. In the long run, any resource that cannot be replenished will run out, and fossil fuels are not being replenished except over long geological time frames.

The basic solar energy resource will not run out. The sun has reliably risen every day for the 5-billion-year history of our planet, and is expected to do so for another 5 billion years. Solar energy has sustained life since it began on our planet. It would be fair to say that a power source that has already worked for billions of years has a proven track record of reliability.

This means that whatever energy systems are put in place that rely on solar energy will not have to be replaced with another energy source in ten, twenty, or a hundred years. If we manage to take the solar step, then we'll have a permanent energy system. Of course, individual pieces of solar equipment may need replacement, but the energy system as a whole can remain solar-based indefinitely. In the 2008 movie *Wall·E*, an autonomous trash-collecting robot keeps itself running for 700 years after civilization has departed by charging its batteries with PV panels, replacing parts as needed. Fictional, but possible.

15. Solar PV Is a Solid Investment for Business

For fifty years after World War II ended, businesses could rely on steady energy supplies and prices. Energy costs were so low that most businesses paid little attention to energy efficiency, and business behavior was not much influenced by energy issues.

We are now entering a period of energy price volatility, so businesses must consider where their energy comes from and how much it costs. Uncertainty about fossil-fuel energy supplies can hinder economic investment, but there is no uncertainty about the energy a solar PV system can generate. Any business that deploys a solar PV system with energy storage will enjoy a reliable energy supply. Solar PV systems have long warranties, require little maintenance, and have no ongoing fuel costs, all attractive features for business investment.

More important, the price of solar electricity does not rise over time, because it is determined by the up-front cost of the PV equipment, not the current volatile price of oil. A business can amortize the known cost for the equipment over the firmly predicted kilowatt-hour energy output of the PV system to enable the business to establish its future energy costs with certainty. With essentially zero economic risk, that amounts to a guaranteed return on investment, a rare thing in any business.

For businesses seeking to reduce operating costs, a capital investment in solar PV will offset annual electricity costs. Once the PV system is paid off, it will continue to generate free electricity for many more years. A solar-powered business has an edge over any competitors that are coping with a future of higher energy prices.

16. Solar PV Is Good for Local Economies

Wherever solar electricity is installed, it boosts the local economy. Most communities import their energy from oil companies and distant utility companies. Money spent on such energy leaves the local economy instead of recirculating to local businesses. The export of money to pay for imported energy acts as a constant drain on a local economy.

When energy is generated locally using solar PV, that money drain is plugged. The money that would have paid for imported energy can instead be spent on local goods and services. Since the money stays local, it can be respent in the community, effectively multiplying its value to the local economy. The Hawai'i Clean Energy Initiative, which hopes to achieve 70 percent clean energy by 2030, has explicitly stated this as a goal. "This will help Hawai'i become more economically stable by keeping an estimated $5.1 billion in state that would otherwise go toward foreign oil investments."[24]

While the PV equipment will likely need to be imported to a local community, it pays itself off after a few years. All energy generated after that period is free of exported money.

Solar electricity puts a local community in the business of *primary production*. In economics, primary production is any economic activity that makes a product from raw natural resources as in mining, forestry, and agriculture. Solar PV qualifies because it converts natural sunshine to usable electricity. Primary production is the foundation of any economy, so installing a lot of solar PV can build a steady base for a local economy.

Solar PV enables more entities to make money by producing energy. Individuals, families, schools, local governments, churches, small businesses, and farmers can install PV and produce electricity as valuable as that produced by a utility company. For many, having a steady income instead of a steady expense from their energy system will help them survive economically.

Solar PV also supplies local jobs for PV installers, since installation tasks cannot be outsourced to other countries. The same can be said for energy-efficiency improvements. No ongoing labor is needed for harvesting the energy over time, but the transition is huge, lasting decades and covering the entire working life of an individual installer. Studies have shown that solar PV generates far more jobs per unit of energy than the fossil-fuel industry.[25]

17. Solar PV Is Good for National Economies

Solar PV helps national economies in several ways:

- Replacing oil with solar reduces oil imports, thereby reducing trade deficits that can drain a national economy. The United States saw $265 billion flow out of the country in 2010 to pay for petroleum imports (which amounts to over half of the $500 billion total US trade deficit).[26]
- Solar is not subject to oil-supply shortfalls caused by political turmoil.
- Solar is not subject to energy-price volatility.
- Solar PV can add primary production to any national economy, not just those lucky enough to have fossil-fuel resources.
- Solar equipment manufacturing, which can take place almost anywhere, can expand a country's industrial base.
- For developing countries, an energy base can be developed locally without dependence on imported energy.

Solar PV brings stability to any national economy that deploys it widely, insulating it from the wrenching effects of oil politics and price speculation. A stable economy promotes confidence in investors, thereby stimulating the investments that sustain economic growth.

18. Solar PV Scales Well

PV systems come in all sizes, from a single panel powering a few lights for a village hut, all the way up to giant solar farms that deploy thousands of panels for utility-scale power generation.

That versatility comes from the modular nature of solar PV. The more panels you install, the more power you produce. The panels are even referred to as "PV modules" in the industry. Modular systems have all these advantages:

- Each home and business can size its PV system to meet its specific energy needs.
- You can start small to offset part of your energy use, and then add more panels later when you can afford them to cover the rest.
- As the population of a region grows, the energy systems can grow with it.

- A gradual transition from fossil fuels to solar can be accomplished without disruption through incremental installations.

Neither clean coal nor nuclear power can scale to different sizes like solar PV. They are economical only when built in extra-large sizes to serve thousands of customers from a centralized plant.

19. Solar PV Requires No Dedicated Land

Most solar PV systems do not need land to be set aside for them. With no moving parts and no noise, solar PV is very flexible in where it can be installed, as long as there is sun exposure. Rooftops are the obvious choice, but panels can also be installed on racks over parking lots, on the ground along highway medians, and on pole mounts on pasture land. The Austrian city of Gleisdorf even erected metal "trees" to support PV panels along city streets.

> In the United States, cities and residences cover about 140 million acres of land. We could supply every kilowatt-hour of our nation's current electricity requirements simply by applying PV to 7 percent of this area—on roofs, on parking lots, along highway walls, on the sides of buildings, and in other dual-use scenarios.
>
> —US Department of Energy[27]

Unlike wind energy, solar PV produces little visual impact. Groups of giant wind turbines with moving blades impose themselves on the visual landscape. Cape Wind, an offshore windfarm slated to be built five miles off the coast of Cape Cod, Massachusetts, was held up for years by coastal residents who thought it would ruin their views of the horizon. Solar PV sits silently in place, attracting no one's attention. PV modules in the form of solar shingles or roof tiles are already available to blend PV completely into a roof.

Newer PV designs could expand the available areas even more. Semitransparent PV modules under development can be tuned to absorb wavelengths that plants cannot use and pass through the wavelengths that plants need for photosynthesis.[28] Such panels could be installed on greenhouses or erected over farmland, providing power without reducing agricultural output. The vast areas of farmland would become available for power production, benefitting the farmer with an additional "crop" of energy.

Figure 39. Oregon Solar Highway installation near Portland, Oregon
Great swaths of unused land next to highways can be used for photovoltaic installations. Revenue from electricity sales could fund highway maintenance.

20. Solar PV Is Ready Now

Solar PV works now. This solution to our energy problem does not need years of research before it can get started. My own home has been operating on solar PV for over fifteen years.

Contrast solar PV's readiness today with clean coal. Although some of the technical steps for clean coal have been demonstrated, there are no full-size clean coal plants operating today. And carbon sequestration is largely unproven. Given that the carbon must be stored without leakage for hundreds of years, sufficient proof will be a long time coming. Building large numbers of clean coal power plants before carbon storage is proven is an enormous risk.

Deploying more nuclear power plants could begin today, but there is not enough natural uranium for a full-scale conversion from fossil fuels to nuclear. That would require breeder reactors, which will require at least a decade of research and development. And in the aftermath of the nuclear meltdowns at Fukushima in 2011, any nuclear construction will be delayed by additional safety requirements and reviews.

Solar PV can begin reducing our carbon emissions immediately for all these reasons:[29]

- Short construction time compared to large centralized power plants.
- Economies of mass production (more like building cars than cathedrals).
- Fewer issues with siting a system.
- Quick decision for individuals compared to big and slow institutions.

The Negatives of Solar PV

Of course, solar PV has negative aspects too, and these must be carefully weighed before we can commit to it.

As an energy source, solar PV has only three disadvantages:

1. Solar PV manufacturing sometimes involves toxic materials.
2. Solar PV relies on good access to the Sun.
3. Solar PV costs are all up front, and perceived as expensive.

The simplicity of installing and using PV panels hides the high tech manufacturing processes used to make them. Those processes often use toxic substances that could be harmful if released into the environment. But methods already exist from the computer chip industry to contain and recycle such materials to minimize potential pollution.

The Silicon Valley Toxics Coalition[30] tracks these issues and provides guidelines for solar PV manufacturers to minimize their toxic releases. Most established PV manufacturers have in-house waste treatment systems to recycle or destroy toxic materials. Those that do not are required by state and federal laws to transport their waste to a toxic-waste disposal facility.

Some PV panels incorporate cadmium, a toxic heavy metal. The cadmium is locked into the panel and does not leach out into the environment while the panel operates. But if such a panel is dumped into a landfill at the end of its life, then the cadmium could leak out over time. That's why recycling of old PV panels is already being set up. The PV CYCLE Association in Belgium accepts old PV panels from any European country and recycles them into usable products.[31]

If toxics are controlled at the point of manufacturing and at the point of disposal, then a PV panel is clean throughout its complete lifecycle because it emits nothing during operation. Compare that to a coal plant, which routinely spews mercury, cadmium and other toxics into the atmosphere on a daily basis, or a nuclear plant that generates dangerous radioactive waste for each kilowatt-hour of electricity generated.

Yet for solar panels to produce clean electricity over their lifetime, they must have access to the sun. One potential disadvantage of solar PV arises when access to sunlight is cut off or reduced, thereby limiting the output of the panel.

How could the sunlight be reduced? The greatest risk comes from a neighboring property owner erecting a building tall enough to permanently shade your PV panels, essentially cutting off the return on your investment in them. Vegetation poses another risk. Over the decades-long lifetime of the panels, trees can grow tall enough to shade collectors. Such factors need to be taken into consideration when siting panels.

A worried PV owner can arrange a *solar easement* with a neighbor. In this private legal contract, the neighbor agrees to not shade the solar collectors, usually in exchange for some form of compensation. Newly built subdivisions may incorporate such solar easements in the initial property titles.

Laws to control solar access can help protect a solar investment, but they are not widespread or comprehensive. California's Public Resources Code (25980) includes the Solar Shade Control Act of 1978, which provides that a tree or shrub cannot cast a shadow that covers more than 10 percent of a solar collector's absorption area at any one time between the hours of 10:00 a.m. and 2:00 p.m., but the law only applies if the tree or shrub was planted *after* the installation of the solar collector. The Solar America Board for Codes and Standards, funded by the US

Department of Energy, has put forth a more comprehensive model statute for states to adopt to protect solar access.[32] So far, only a few states and cities have adopted parts of the statute.

Sunlight can be blocked on a global scale through large volcanic eruptions. The effluvia injected into the atmosphere by the 1991 eruption of Mount Pinatubo in the Philippines cut solar energy worldwide by 10 percent, but the effect lasted only a couple of years.

A potentially longer-term risk comes from increases in global cloud cover due to global warming. Extra heat means more evaporation of water from the oceans, which might produce more clouds—unless the higher temperatures prevent clouds from forming. The computer modeling of climate cannot yet predict whether global cloud cover will increase or decrease.

Geoengineering could also cut solar input. There are proposals to solve the problem of global warming by reducing the amount of sunlight reaching the Earth's surface. These projects would alter the Earth's atmosphere to reflect more sunlight back into space, thereby cooling the planet.

Natural or human-induced reductions in global solar input are not likely to be large enough to prevent PV from working. We can compensate for any reductions in energy output by installing more collector area, though this will make PV a bit more expensive.

This brings us to the last disadvantage of solar PV—the initial cost of setting it up. The cost structure of solar PV differs so completely from that of fossil fuels that most people perceive solar PV as expensive. This widely held belief stands as the single most significant barrier preventing our shift to solar energy, so I've devoted the next chapter to that subject.

CHAPTER 16

The Compelling Economics of Solar PV

Despite all of the advantages described in the previous chapter, solar PV produced less than 1 percent of the world's electricity in 2012. All the good features of solar are having little effect because the numbers are so low. Why?

The main reason is that solar PV got a late start. Fossil fuels have been growing for over 200 years, while photovoltaics date from the 1960s. And during the first two decades of PV, prices were prohibitively high, restricting PV to specialized locations such as remote mountaintops and orbiting satellites.

Even fifteen years ago, when I purchased my first system, PV electricity was expensive relative to utility-produced electricity. In our case, the high cost of stringing a power line to our rural property made solar PV competitive at that time, even with the added cost of batteries and a backup generator.

Now the prices of PV panels have dropped to a tenth or less of what they were then, making the decision to go solar much easier for many more people.

The basic economic proposition of solar PV is simple: Solar energy is free, but the equipment to harvest it is not. With solar, you're not buying energy, you're buying hardware to produce energy. That hardware is tangible and long lasting.

The basic economic proposition of fossil-fuel energy is even simpler: You buy energy, you consume energy, you buy more energy. The energy is ephemeral, dispersing as tepid heat into the surroundings when you're done with it, and you're left with nothing but a need for more energy.

This fundamental difference makes it tricky to compare the cost of solar PV electricity to utility electricity. With solar PV, you pay all the money up front to install a PV system, and then you harvest free energy for the lifetime of the system, with only minor outlays for maintenance. With utility electricity, there is no installation cost for you, but you pay as you go for the electricity you use, year after year. The only fair comparison would be over the period of time that the PV panels produce electricity, typically taken to be thirty years for well-made panels.

But even then, the cost comparison depends on where you are located:

- Different regions receive different amounts of sunshine, so a given solar panel will produce different amounts of energy in different regions.
- Each individual PV site has is own combination of roof arrangement, mounting angles, and potential shading at various times of the day or year. These affect how much electricity is produced by the panels.
- Different regions have different costs for grid power. Most have price tiers to encourage energy conservation, so you pay a relatively low price per kilowatt-hour for the first quota of kwh, and then higher prices as you use more. Many regions also have time-of-use rates that charge more for midday peak electricity and less for off-peak use. Some also charge more in summer when air-conditioning loads are high.
- Different regions have different government incentives for installing solar PV.

Because of all these differences, no single cost comparison applies equally everywhere. Each PV system is custom designed for its location and requires a bit of research to determine its economic feasibility.

Today such due diligence is usually performed by the local solar contractor who is proposing to install a PV system. They will know the current price of the hardware and installation costs, your local utility rates, and the incentives that apply in your area. If they don't, find another contractor, or do the research yourself.

You could start with the PVWatts solar calculator available online through the US National Renewable Energy Laboratory.[1] You enter a location anywhere in the world, the local average utility rate, and the basic characteristics of the proposed PV system. It will then calculate the amount and value of the solar electricity generated for each month, and a total for a year.

But this simple calculator does not take into account tiered pricing for utility electricity, which can make a big difference. For example, the utility company at my location in northern California charges about 16¢ per kilowatt-hour for the first 200 kilowatt-hours per month (summer residential rates in 2015). Every kwh over 400 per month will cost you 34¢, over twice as much. A solar PV system will shave off electricity from the top tier first, helping to pay for the system much faster than at the baseline rate. Your utility bill should tell you how much of your electricity is over the baseline amount.

Similarly, time-of-use rates can also make a big difference for those who have that service. A time-of-use electric meter includes an internal clock that records when each kilowatt-hour was used. In our territory, peak hours are defined as noon to 6:00 pm, Monday through Friday, while all other hours are considered off-peak. The local off-peak residential baseline rate is 11¢ per kwh, while the on-peak summer baseline rate is 36¢ per kwh. A solar PV system happens to maximize its output during the peak hours, making it particularly effective at displacing the most expensive utility electricity.

Since the economic value of a solar PV system is highly individual, no one can make a universal declaration that solar is cost effective or not. You have to run your own numbers. Many solar contractors have computer programs loaded with local utility rates and solar data, and can generate a readout for your specific system in your location.

A big uncertainty lurks in all comparisons of solar to utility power: How much will utility electricity cost in the future? You could assume the price will remain the same in future years, but that would be overly optimistic. You have to pay as you go, covering whatever the utility company charges in future years. Those charges have shown a pattern of gradual increases over the years. If and when the costs of controlling carbon emissions are included in electricity rates, the price for electricity from fossil fuels will definitely go up. Any such utility rate increases favor solar PV.

Note that the uncertainties do *not* come from the solar PV side of the comparison. The cost of PV electricity can be precisely determined. The purchase price of a given PV system is an indisputable fact known at the time of installation, and the number of kilowatt-hours produced by that PV system is entirely predictable based on location and on averaging over many years. That certainty on the solar side permits a simple calculation to determine a firm price for your solar electricity.

Here's how my own PV system works out. I have 3,800 watts of installed PV capacity. If I paid the current price of $3 per watt for an installed system ($4 per watt installed, minus subsidies), that would be $11,400. With that PV system, I generate an average of 4,800 kilowatt-hours of electricity per year in our location. Over the thirty-year lifetime of the system, that totals to 144,000 kwh. If I divide the total cost of the system by the total kwh produced by it, the result is 7.9¢ per kwh over the lifetime of the system. According to the website of my local utility, the average electricity rate for low users (like us) is currently 17.9¢ per kwh.[2]

So my PV cost of 7.9¢ per kilowatt-hour is *half* the current utility rate of 17.9¢, when averaged over the lifetime of the system. And if utility rates go up as expected, the comparison gets better over time.

Your PV rate comparison will not be the same as mine. To figure your own PV electricity rate, you need two facts: the total cost of your PV system, and the estimated electricity output of your PV system in your region over the lifetime of the system. Then you can compare that to the average electricity rate from your local utility. A local installer of PV systems should be able to provide all that information.

If your solar PV rate is lower than your local utility's current rate, then you know you have a solid long-term investment. You may pay a sizable sum up front, but your electricity will be cheaper in the long run. For the price of a new car, you get a supply of clean electricity that will last a generation, at a price that is locked in for thirty years.

Of course, I may not occupy my house for the full thirty years, but I won't lose money because solar PV adds value to a home. A study at the Lawrence Berkeley National Laboratory found that "homes with solar photovoltaic (PV) systems sell for a premium over homes without solar systems."[3] Another study by the National Bureau of Economic Research found that solar PV adds between 3 percent and 4 percent to the selling price of a home.[4] The added value is roughly comparable to the cost of installing the system. So I can recoup the value of my system when I sell our home, and invest the money in PV for our next home.

Long-lived institutions like schools and hospitals can get the full thirty-year benefit of a solar PV installation. Businesses should be especially attracted to investing in solar PV for the long-term guarantee of a low and stable electricity rate.

An economist might consider this view of solar PV economics oversimplified. Conventional economic calculations would include the cost of borrowing money to install the system, the depreciation of the system, and the discounting of future energy gains to assign them a "net present value."

But conventional economic analysis of energy is also guilty of oversimplification, by omitting many costs and benefits of different kinds of energy. Economists treat energy as just another commodity. From that point of view, all electricity is the same, regardless of how it is generated. But if you widen your energy scope, you find that different sources of electricity have different consequences that are not accounted for in the price of the electricity.

Fossil fuels have many *hidden costs*, meaning costs not included in the price, and paid for by someone other than the buyer. For example, our fossil-fuel energy systems produce air pollution in the form of hydrocarbons, ozone, particulates, carbon monoxide, nitrogen oxides, and sulfur oxides. These pollutants create health problems and damage crops and forests, generating costs that are not paid for by the producer or the consumer of fossil-fuel energy. Since the vendors of fossil fuels usually do not pay anything for these negative effects, the price of their product does not reflect the complete set of costs associated with using their product. That makes air pollution a hidden cost of fossil fuels. There are many others, as described in the section titled "Ten Reasons Why Fossil Fuels Are Not Good For Us Anymore" (page 81).

In a pure market-driven economy, hidden costs should not exist. Economic decisions about which energy source to use should weigh all the costs and benefits of the choices, so we can pick the ones with the lowest costs and greatest benefits. If some of the costs and benefits are left out of the calculation, then the economic conclusion cannot be accurate.

The problem is that these hidden costs are enormously difficult to compute. It's not possible to prove that *this* bit of fossil-fuel air pollution caused *that* child's asthma attack, requiring an expensive visit to the emergency room. Rather, the air pollution spills into the atmosphere and spreads out over a wide area, mixing with other sources of pollution. The best one can do is to measure how each increment of air pollution affects a controlled study group, and then extrapolate that to the entire exposed population.

The hidden costs of carbon dioxide emissions are even harder to calculate. Because carbon dioxide persists in the atmosphere for decades, it gradually spreads out over the entire surface of the planet. Each bit of carbon dioxide contributes to global problems, but no one can be assigned responsibility because almost everyone uses fossil fuels. So every bit of fossil-fuel carbon dioxide becomes everyone's problem, but at the same time, no one's problem.

That's the situation we face today in choosing our energy sources. Many of the costs of fossil fuels are hidden, and many of the benefits of solar energy are ignored, so the economic picture is distorted. If you zoom out your energy scope to encompass the full picture of our energy choices, then we can at least note the hidden costs even if we cannot compute them.

For example, solar PV doesn't produce carbon dioxide during its operation. In judging the value of solar electricity, it would be more fair to compare it to fossil fuels used in a system with 100 percent carbon capture and permanent sequestering. Since such complete carbon capturing does not yet exist, those costs are unknown.

We could try comparing the cost of solar PV to the cost of climate change and ocean acidification, which are the consequences of continuing with fossil fuels. But again, such costs are difficult to compute with any accuracy.

The *Stern Review on the Economics of Climate Change* [5] prepared for the British government in 2006 by economist Nicholas Stern, chair of the Grantham Research Institute on Climate Change and the Environment at the London School of Economics, attempted to assign specific costs to the effects of climate change. The report concluded that the benefits of strong early action on controlling climate change will outweigh the costs of such action, but the impact of the study was lessened because it still had to make many assumptions to arrive at projected costs. Those assumptions were subject to challenge by climate-change skeptics and other economists.[6]

A study by the US National Research Council titled *Hidden Costs of Energy: Unpriced Consequences of Energy Production and Use*[7] would appear to provide some answers. While noting many hidden costs of fossil fuels, its authors admit to being unable to assign monetary value (to *monetize*, to use their term) to the costs. The study's authors found they could assign actual costs to only one hidden cost—air pollution. Here's the opening paragraph of their conclusions:

> In aggregate, the damage estimates presented in this report for various external effects are substantial. Just the damages from external effects the committee was able to quantify add up to more than $120 billion for the year 2005. Although large uncertainties are associated with the committee's estimates, there is little doubt that this aggregate total substantially underestimates the damages, because it does not include many other kinds of damages that could not be quantified for reasons explained in the report, such as damages related to some pollutants, climate change, ecosystems, infrastructure, and security.

A Harvard study that looked at the hidden costs of coal energy in the eastern United States concluded that those costs amount to $345 billion per year, which would add 18¢ to each kilowatt-hour of electricity generated from coal. And the authors acknowledge that they did not include many other hidden costs that they could not quantify.[8]

For oil, a major hidden cost comes from the military efforts by the US government to protect the world oil system. Here the difficulty comes from separating those expenses from the normal global operations of the US military. Three recent estimates of the cost to taxpayers of defending oil in the Middle East ranged from $29 to $143 billion per year.[9] The wide range shows the difficulty of assigning specific costs to that mission. Of course, oil companies do not cover any of that cost, so it is not reflected in the price of imported oil.

Also missing from the other side of the energy ledger are the many benefits of solar PV. The previous chapter enumerated twenty such benefits, including safety, reliability, economic stability, and national security. Each of those benefits adds value to solar PV, value that is not present in electricity from fossil fuels.

Like the hidden costs of fossil fuels, the hidden benefits of solar PV cannot be easily monetized. Most of these benefits accrue not to the individual using the electricity but to society in general. The fact that my solar PV system does not contribute to climate change has no direct economic effect on me personally. There is no way to accurately compute what global warming costs were avoided by my small increment of solar electricity.

So economists, unable to assign numbers to these extra values, leave them out of energy-cost comparisons. That leaves solar PV undervalued when we make decisions about our energy future. Much of the value of solar energy is missed by traditional economic analysis, according to solar economist Richard Perez of the University at Albany, State University of New York.[10] As the Executive Summary of the Stern Review states: "Climate change presents a unique challenge for economics: it is the greatest and widest-ranging market failure ever seen."

But just because we cannot assign a number to a value doesn't mean it's worthless. If the economic calculations have to pretend that these very real and important solar benefits do not exist, and that the very real and important hidden costs of fossil fuels do not exist, then the conclusions derived from those calculations cannot reflect reality. Choosing energy based solely on price got us into our current energy mess.

Fossil fuels remain cheap because they're dirty. If fossil-fuel energy systems had to clean up all the messes they make (climate change, ocean acidification, air pollution, water pollution, etc.), then fossil fuels would no longer be cheap. The Harvard study cited above indicates that the price of coal electricity would double or triple if it included coal's hidden costs.

The problem is that while all members of society share the benefits of solar PV, the cost of a solar PV system falls to an individual. Most individuals are not in a position to make sizable investments for the good of humankind.

Since our current market system cannot assign value when faced with real but incalculable costs and benefits, governments have stepped in to correct the imbalance. When a government provides you a tax rebate or other financial subsidy for solar PV, that economic support is an attempt to pass some of the social savings on to the individual making the investment. You can act in your own interest and yet benefit society.

In a sense, society becomes a silent partner investing in your energy system. Society's return on its investment comes in the form of a stable and clean energy system to power the economy. Society certainly does get something for its money, even if it's difficult to monetize.[11] So when you run the numbers for your proposed solar PV system, be sure to include any such help.

You might think that such rebates distort the financial picture, unfairly favoring solar over fossil fuels. But as I describe in chapter 24, *What Holds Us Back?*, fossil fuels are massively subsidized around the world on a scale far larger than any solar rebates. That economic distortion currently favors fossil fuels, not solar. Given the long list of problems from fossil fuels, their subsidies are of dubious value to society, and many people consider them counterproductive.

Most of the government subsidies for solar have been temporary, to boost the initial phase of a new industry until it can stand on its own. Some wonder if the current government support for solar were to be removed, whether the solar PV industry would collapse, as the solar water heating industry did in the 1980s when its subsidies were rescinded. The next chapter shows why that's not likely, for the simple reason that solar PV prices are on a downward trajectory, and will soon not need any subsidies.

PART III: SOLAR IS THE ONE

CHAPTER 17

The Price of Solar PV is Dropping

The price of solar PV will soon change from a hindrance to a help. The price is dropping, while the prices of fossil fuels are rising. If those trends continue as expected, then at some point the prices will cross over, and solar PV will be cheaper than fossil fuels. When that happens, solar PV will take off, driven entirely by market forces as the cheapest energy available.

The price of solar PV has been going down for several decades. The first generation of solar PV cells from the 1960s were made of single-crystal silicon, primarily for NASA for powering satellites, where reliability requirements far outweighed cost considerations. PV panels were handmade to exacting standards for such missions, and so were very expensive, over $100 per watt. Only NASA could afford to spend $10,000 to light a 100-watt bulb.

Back on Earth, the need for electricity at remote sites led to the development of a terrestrial PV market. For example, a telephone company might place a microwave relay tower on top of a mountain to transmit calls over rugged terrain. In the past, powering that tower required a diesel generator and periodic trips up the mountain to deliver fuel. A solar PV system with batteries operates reliably without the expensive diesel deliveries, making it cheaper than the generator system.

The Swanson Effect

Price of crystalline silicon photovoltaic cells, in dollars per watt

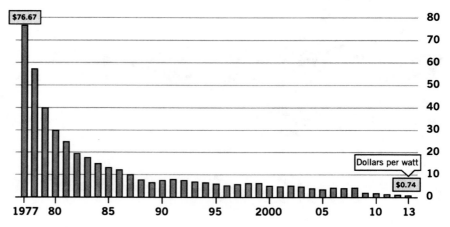

Figure 40. Historical prices of PV[3]

The general trend for photovoltaic prices continues downward as manufacturing methods improve in efficiency and scale up in size. Each price drop creates more demand, further spurring innovation.

By 1980, the retail price of PV had dropped to $22 per watt, and by 2010 to $1.50. In 2009, panels from Arizona-based First Solar, Inc. dropped below $1.00 per watt in manufacturing cost (not retail price) for the first time. The average price of PV fell another 50 percent in 2011[1] and continues to go down.

Much of the drop in prices has come from economies of scale from mass production. In the early days, PV experienced the classic "chicken and egg" problem for its market—prices were high because demand was low, but demand was low because prices were high. As demand gradually rose over the decades, production volumes rose, justifying investments in automation to bring the price down further.

This trend even has a name: Swanson's Law, named after Richard Swanson, the founder of SunPower, a big solar-cell manufacturer. He noted that the cost of photovoltaic cells falls by 20 percent with each doubling of global manufacturing capacity.[2] This graph shows the dramatic drop in PV prices over several decades.

Further price reductions will likely come from new PV technologies. First Solar, Inc. achieved its record low price by using *thin-film PV.*

Instead of growing solid silicon ingots and slicing them into wafers to make solar cells, the thin-film process deposits extremely thin layers of active material on an inert substrate like glass. By using the least amount of expensive material, thin-film PV can reach unprecedented low prices.

Now demand is high enough to stimulate research into many new technologies for making PV. Any new PV technology such as thin film must achieve three goals simultaneously: an efficiency of at least 10 percent, durability to last decades exposed to sun and weather, and low cost. Many can achieve one or two of these goals, but a technology must reach all three to be commercially viable.[4]

Here are some of the new PV technologies being developed:

- *Dye-sensitized PV cells* employ a thin layer of dye material on titanium dioxide with nano pores to add surface area, and with a liquid electrolyte flowing behind to supply electrons to the dye. Top efficiency in the laboratory has reached 10.4%, but it uses rare metals, and the liquid electrolyte could leak.[5]

- *Biomimetic PV cells* mimic the active center of biomolecules such as chlorophyll and hemoglobin. Organic molecules such as porphyrins and phthalocyanines can have their energy levels tuned. Stability over time is an issue, as is efficiency.[6]

- *Organic PV cells* replace expensive silicon semiconductor material with cheaper polymer semiconductors. Top lab efficiency has reach 7.9%, but the plastic material degrades over time, and needs to achieve higher efficiencies.[7]

- *Perovskite PV cells* lay down a thin film of perovskite, an inexpensive natural mineral composed of calcium, titanium, and oxygen. Efficiencies as high as 15% have been recorded.[8]

- *Multi-junction PV cells*, where different layers are optimized to absorb certain wavelengths while letting other wavelengths pass through to layers below. Maximum theoretical efficiency is 87%, over twice that of single-junction PV (34%). Actual efficiencies in the lab have reached 42%, and commercially available cells have 30% efficiency, but the complexity leads to higher manufacturing costs.[9]

- *Cast silicon PV cells*, where extremely thin crystalline silicon cells are formed directly in a mold instead of being cut from an ingot, which wastes up to half the material as sawdust. The company, 1366 Technologies, Inc., has demonstrated 17% efficiency[10] in its cast cells. In 2013, the company opened a demonstration-scale factory in Bedford, Massachusetts.[11]

- *Plasmonic PV cells* use a phenomenon called surface plasmon resonance to improve the efficiency of thin film PV. Most thin film PV cells have trouble absorbing sunlight because they have so little material to interact with the light. These cells employ special metal nano particles to trap light long enough for it to be absorbed before it passes through. [12]

- *Quantum dot PV cells* are made up of particles of PV material so small that their quantum energy levels can be tuned to absorb particular wavelengths. In particular, they can be tuned to infrared, which other PV cells cannot convert. Half of the solar radiation reaching the earth's surface is in the infrared range. Their maximum theoretical efficiency is 65%, but as of 2012 they've reached only 7.0% in the lab.[13]

- *Solar nantenna (nano antenna) collectors* operate on a completely different principle than PV cells, which rely on discrete energy levels tuned to specific light frequencies. Nantenna solar panels behave more like a broadband radio receiver. They use nanoscale light-sensitive antennas and high-speed electronic diodes to convert the electromagnetic light waves directly to electric current. Currently they can convert only lower-frequency infrared light, but that would include infrared radiated from the earth's surface at night. Higher frequencies in the range of visible light will require further refinement of the technology. A broadband panel could be aimed up during the day to collect direct solar radiation, and aimed down at night to harvest infrared. Theoretical maximum efficiency is 85%, and the expected overall system efficiency is 46%.[14] University of Connecticut Associate Professor Brian Willis has patented a novel fabrication technique that could make solar nantennas feasible to produce.[15]

With conventional PV prices continuing to decline due to mass production, and with many new PV technologies promising to bring even lower prices, it's safe to project a continuing reduction in the cost of PV.

At the same time, fossil-fuel prices are expected to go up to pay for the costs of global warming.

Putting those two trends together predicts a coming crossover, when the cost of solar PV electricity becomes less than the cost of fossil-fuel electricity.

The crossover will not be a distinct point in time. There will be no headline declaring SOLAR PV NOW CHEAPER THAN FOSSIL FUELS, because there is no single price to compare. Fossil-fuel prices differ by region, and those prices swing up and down based on demand, market speculation, and political developments. At the same time, the price of electricity from a solar PV installation varies with location, system size, and financing mechanism.

Solar PV is already cheaper than utility electricity in sunny locations that have high utility rates. As PV prices continue to go down, more locations with less sun and lower utility rates will be brought into the fold, allowing PV to gradually spread over the surface of the planet.

As that happens, the shift from fossil fuels to solar PV will accelerate. If we can gain all the previously described benefits of solar PV *and* get them at a lower price, there will be no reason to stay with fossil fuels.

Eventually the subsidies for solar energy can be phased out. Solar PV will not need international treaties, carbon-trading markets or tax incentives, as individuals and organizations will choose it simply to save money. Future solar PV will be driven by market demand.

PART III: SOLAR IS THE ONE

PART IV
A Plan

CHAPTER 18

How Much Solar is Needed?

To halt climate change and ocean acidification, we need to shift our goal upward from *reducing* fossil fuel use to *eliminating* fossil fuel use.

> The temperature at which global warming will finally stop depends primarily on the total amount of CO_2 released to the atmosphere since industrialization. This is again due to the long life-time of atmospheric CO_2. Therefore if global warming is to be stopped, global CO_2 emissions must eventually decline to zero.
>
> —The Copenhagen Diagnosis[1]

Pushing carbon dioxide emissions to zero requires replacing all fossil fuels with CO2-free energy sources. With its many benefits and its trend of declining prices, solar PV is poised to grow into a major source of energy for humankind, but can it really get big enough to make fossil fuels obsolete?

Any new source of energy faces a daunting challenge—humans use a lot of energy. So any new source must also be able to produce a lot of energy. The current rate of energy use in the world averages 15 terawatts (15 trillion joules per second), of which 85 percent comes from fossil fuels.

The challenge is even greater for an emerging energy source. It must:

- Replace existing fossil fuel energy use.
- Extend modern energy services to the 1 billion people on the planet without them.
- Grow as the world population grows in the future.

Certainly the basic solar resource is big enough. Our fossil-fuel operations are massive, but they are dwarfed by the sun power reaching the surface of the Earth, which is 6,000 times larger than the rate of all human energy use on the planet. Every five days, the sun delivers the energy equivalent of all the fossil-fuel reserves in the world. University of California economist Stephen J. DeCanio says that solar power is the only non-fossil fuel energy source big enough to meet future world energy needs in an environmentally benign way.[2]

Several studies have estimated how much solar equipment would be needed. A 2011 study by the Institute for Policy Research and Development (IPRD) assumed a modest 30 percent reduction in the rate at which people use energy in developed countries such as the United States, Japan, and Europe. They then extended that rate of energy use to everyone on the planet, equalizing the developed and developing worlds. They concluded that such a world could meet its energy needs entirely from wind and solar in a few decades.[3]

Since we must use the current energy system to build the next one, the IPRD study also examined the energy burden of building a complete solar-based energy system. They estimated that by dedicating just 1 percent of current fossil-fuel energy use to the transition, we can build the wind and PV to convert the entire world energy system in forty years, using existing solar and wind technology. The transition becomes self-sustaining as solar takes over, if about 10 percent of the solar capacity is dedicated to completing it.

Another study by Mark Jacobson of Stanford University and Mark Delucchi of the University of California at Davis showed that a mix of solar-based technologies, including 19 terawatts (TW) of wind turbines and 17 TW of solar PV, could meet all energy needs by 2030 or 2050 depending on how aggressive the transition is pursued. They concluded that obstructions to such plans are mostly political and social, not economic or technological.[4]

Location	Average rate of energy use per person (2011), in watts (joules per second)
United States	9401
Germany	4962
Japan	4976
China	2332
Brazil	1845
India	593
Ethiopia	49
World average	2253

Figure 41. Worldwide variations in energy use[8]

Each number divides the total energy used in a country for all purposes by the population of that country. The numbers include industrial, commercial, and transportation energy use in addition to personal energy use. They indicate how much energy an entire society uses, per person.

A Greenpeace study's Advanced Energy scenario projected total world energy flow of about 15 TW by 2050, with renewable energy sources providing 80 percent of that.[5] A study from the Union of Concerned Scientists estimated energy use of 15.4 TW by 2030, of which 50 percent could come from solar sources.[6]

Each of these studies has to make certain assumptions, from which it draws its conclusions. It is possible to build our own rough estimate from the bottom up, making some basic assumptions:

- World population stabilizes at around 10 billion people in 2060.[7]
- Everyone in the world has access to modern energy services.
- Everyone has sufficient energy to meet their needs, without being wasteful.

The first two assumptions are pretty solid, but the last one requires more care. How much energy does one need if we are not wasteful?

Today, rates of energy use vary greatly among different populations, as the following figure shows.

We can see that Germany and Japan use energy at the average rate of about 5 kilowatts per person for a modern lifestyle. That rate could be reduced by 30 percent through ordinary energy-efficiency measures, including retrofitting existing buildings or replacing them with more efficient buildings over several decades.

We can gain another 30 percent reduction simply be replacing heat engines with electric motors. Most fossil fuels are burned in heat engines, which discard 65 percent to 75 percent of the input energy as waste heat. Since heat engines are used for most electricity production and almost all transportation, it is estimated that 53 percent of energy used worldwide is thrown away as waste heat.[9]

Replacing heat engines with electric motors powered from solar PV eliminates that wasted energy. The solar output only has to match the energy *output* of the fossil-fueled heat engines, not the input. Because solar PV generates electricity directly, there is no need to replace the 65 percent to 75 percent of fossil-fuel energy that ends up as waste heat from heat engines. All the studies cited above make this adjustment, and specifically the IPRD study conservatively estimated the savings at 30 percent, so we will use that number here.[10]

So by cutting 30 percent on the consumption side with efficiency improvements, and by cutting another 30 percent on the production side by switching to an all-electric economy, we can achieve a total of 60 percent reduction in energy use. That would let us reduce the 5 kilowatt rate of energy use per person to 2 kilowatts, without sacrificing quality of life for those in developed countries. And we can extend that 2 kilowatts per person to all those in the world without modern energy today to raise their standard of living.

So by 2060 we would need 20 TW of total energy flow to supply 2 kilowatts on average to each of 10 billion people. That's larger than the 15 TW used today because this estimate includes population growth as well as extending energy services to more people. But it is not as large as it would have been if we simply extended today's inefficient fossil fuel energy systems to everyone, an entirely unreasonable prospect. The 20 TW estimate matches that of a 2007 study that said stabilizing carbon dioxide at climate-safe levels will require 15–20 TW of carbon-free power by 2050.[11]

So if the goal is roughly 20 TW of usable energy flow by 2060, how much solar equipment would that take? All the studies cited above agree that we will most likely see a mix of solar-based technologies, including solar PV, solar thermal, wind, hydroelectricity, and biofuels. Each study assembles a different mix, based on its own assumptions.

But you must not assume that all these solar energy sources are the same and will contribute equally in the future. Many will reach inherent limitations before reaching 20 TW. For example, electricity from wind turbines is competitive with fossil fuels today, but requires special sites that have sufficient wind and few people. High-quality wind energy is not nearly as ubiquitous as solar energy.

Conventional hydropower is clean and efficient, but requires large land areas for lakes behind hydroelectric dams, and such floodings have their own problems. The area required for 100 MW of hydropower is about 12,000 acres, while the area for a 100 MW solar PV farm is only about 800 acres.[12]

Biofuels on the scale needed to make a dent in fossil-fuel use will compete with food production for land and water. The net energy productivity of all biological growth on the planet is only 55–60 TW, and humans already appropriate 30–40 percent of that for food, feed, fiber, and fuel (mostly wood for cooking in villages).[13] Doubling that to meet our future energy needs would leave too little for nature to continue to function. "Proposals for massive biomass energy schemes are among the most regrettable examples of wishful thinking and ignorance of ecosystemic realities," says Vaclav Smil, Distinguished Professor Emeritus in the Faculty of Environment at the University of Manitoba in Winnipeg, Canada.[14]

Solar PV stands out from the rest of the solar-based energy sources for at least four reasons:

- Solar PV uses direct and indirect sunlight, so almost any location has the raw resource available.
- Solar PV can be sited in the midst of populated areas because it has no moving parts, makes no sound, and can be placed on rooftops and above paved areas.
- Solar PV has great potential for reaching the lowest price.
- Only solar PV scales relatively harmlessly to huge numbers and can absorb all the energy-poor of the world.

How much PV would be needed? Let's assume solar PV provides 75 percent of the future 20 TW energy needs, with a mix of wind, hydroelectric, solar thermal, and biofuels supplying the remaining 25 percent. So PV needs to provide 15 TW on average.

Efficiency vs. Capacity Factor

The *efficiency* of a PV panel measures what percentage of the solar energy hitting the panel is converted to electricity. Efficiency is a characteristic of the panel, determined by the panel's design and materials. The *capacity factor* compares any panel's actual energy output to the energy the panel would produce if it were aimed at the sun twenty-four hours per day. Only a panel orbiting the sun could reach 100 percent capacity factor, while panels on Earth are subject to nighttime and weather. Capacity factor is a characteristic of the latitude and clouds at a given location. Any panel in the same location and mounting angle would have the same capacity factor, independent of its efficiency.

But we cannot just install 15 TW of PV panels and expect to get 15 TW of steady power out of them. That's because the rating on a PV panel is a measure of its *maximum* output in full sun. Because of night and weather, a PV panel typically *averages* 12 to 20 percent of its maximum output, depending on latitude and climate. This measure is called the *capacity factor*, the fraction of its maximum capacity that it actually delivers (see sidebar). A capacity factor of 100 percent means the device is running at maximum output all the time.

The following figure shows the average capacity factor as measured for various utility-scale generators running in the United States.

If we assume an average PV capacity factor of 15 percent for a mix of residential, commercial, and industrial PV, we would need to install 100 TW (peak) of PV panels to deliver 15 TW of average power. How much area is needed for 100 TW of solar PV?

PV panels today are about 15 percent efficient at converting sunlight into electricity (not the same thing as 15 percent capacity factor; see the sidebar). Such panels can produce about 15 watts of peak output for each square foot of panel. If we need 100 TW of PV panels, that amounts to 258,000 square miles of PV panels, which is about the area of Texas, less than half of one percent of the world's land area. If we

Generator Type	Average Capacity Factor
Nuclear	90%
Coal	60%
Wind	32%
Solar Photovoltaic	19%
Solar Thermal	18%
Natural Gas Turbine	4%

Figure 42. Capacity factors for utility-scale generators, United States 2013[15]

Nuclear power plants are run at full output with few interruptions, while coal plants require more maintenance, resulting in more down time. Wind farms are subject to the variability of the wind, and solar energy output is reduced by clouds and nighttime. Natural gas turbines can be higher than 4%, but are typically used intermittently to fill in gaps in supply.

manage to double the efficiency of PV to 30 percent over the coming decades, we would need only 129,000 square miles of panels. That's about the area of New Mexico. We do not need to cover the planet to meet humanity's future energy needs with solar PV.

These numbers are easier to comprehend if scaled down to the personal level. If the future population of 10 billion needs 100 TW of PV, then each person needs 10 kilowatts of PV. Keep in mind that this is not the size of a personal PV system. That 10 kilowatts supplies one person's portion of all the energy flowing in their society, including personal, commercial, industrial, and transportation energy use.

A 10 kw PV system with 15 percent efficient panels would cover an area of 680 square feet, or a square about twenty-six feet on a side. With future 30 percent efficient PV panels, that area could be reduced to 340 square feet, or a square about eighteen feet on a side. These are not unreasonably large areas, considering that they would meet the entire world's energy needs,

Of course, the residents of Texas would object if we tried to cover every square foot of their state with PV to supply the world's energy needs. Instead, that total area will be distributed all over the planet in smaller units, each strategically placed to deliver the electricity where it is needed.

Finding such areas to install PV will generally not be a problem. Some of the electricity will likely come from large-scale PV farms covering many acres. Airports can dedicate some of their large unused buffer areas to solar PV, as can some industrial sites and highway interchanges and medians.

But PV does not require dedicated land. Because solar PV has no moving parts and makes no noise, it can be mounted on or above just about any surface that gets sun. The US Department of Energy estimates that cities and residences in the United States cover about 220,000 square miles.[16] Providing each US resident today with a 10-kilowatt PV system would cover 7,400 square miles with today's PV panels, which would amount to less than 4 percent of that already built-up area.

PV panels can be installed out of the way on home and commercial rooftops. A California study found that about 25 percent of all residential rooftops and 60 percent of commercial rooftops are suitable for PV installations.[17]

Vast areas of existing parking lots could be shaded with elevated racks holding PV panels. For example, the Cincinnati Zoo covered 1,000 parking spaces with solar PV canopies, providing power to the zoo and keeping the parking lot cool in summer. Tall buildings could use PV awnings to generate power and reduce air-conditioning loads. Some buildings are even experimenting with semitransparent PV windows on vertical surfaces that provide views, lighting, and power.[18]

As the PV industry grows, we will see more applications of such building-integrated PV systems that solar-activate part of a building's exterior. When a new roof is installed, PV shingles and PV tiles are available to replace conventional roofing materials. In some cases, specially designed PV panels can directly replace seamed-metal roofing.

Solar PV has even been installed over water. The Far Niente Winery in northern California floats 994 panels mounted on pontoon rafts on their irrigation pond. They chose a pond-mounted system, which they call Floatovoltaic, so they could minimize the land area taken away from grapevines. In London, solar panels form a canopy over the Blackfriars Bridge that carries trains across the River Thames.

Because solar PV is entirely modular, systems can be of any size. That lends great flexibility to siting them, because they can be sized to fit any available area. That makes the transition from fossil fuels to solar much easier. We have only to find enough areas and install panels on them.

We can also reduce the area needed by making PV panels more efficient. If we can go from the current 15 percent efficiency to 30 percent, we would need only half the original area. Continued research in PV technology will most likely bring about such efficiency improvements over time.

Even with efficiency improvements, the shift to solar PV will be a massive undertaking, requiring rapid exponential growth if we are to drive fossil fuel consumption to zero. The next chapter describes a possible growth path for solar PV.

PART IV: A PLAN

CHAPTER 19

Exponential Growth

I t was in 2012 that the installed PV capacity around the world reached the milestone of 100 gigawatts (GW).[1] But that achievement amounts to only one-tenth of 1 percent of the 100 terawatts (TW) of PV capacity needed to replace fossil fuels.

Clearly the task is huge, and seems impossible. During 2014, about 44 GW of solar PV panels were installed. At that rate, it would take 2,200 years to reach 100 TW. By then we will have thoroughly cooked our planet with carbon dioxide if we continue with fossil fuels.

Fortunately, we can achieve the goal in a few decades by accelerating PV production from linear growth to exponential growth. Linear growth means you grow by the same quantity each year. So in 2014 we added 44 GW, in 2015 we add 44 GW, etc. Repeating the 44 GW of output every year into the future amounts to linear growth.

But linear growth assumes that the demand for PV doesn't increase over time. It assumes that we build no new PV factories, so the same PV factories produce the same number of panels every year, with no growth in the industry.

In a more realistic scenario, the PV industry grows each year, expanding the capacity to produce PV panels. That growth is usually presented as a percentage over the previous year's output. For example,

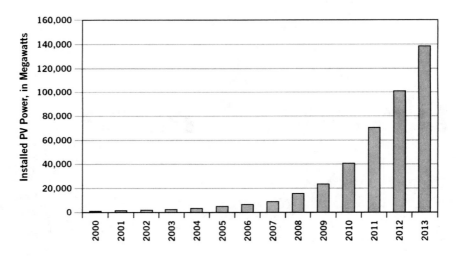

Figure 43. Exponential growth of PV[3]

In an exponential growth pattern, each year's growth is larger than the previous year's growth. The photovoltaic industry is experiencing rapid exponential growth, which will enable it to eventually catch up with fossil fuels.

in 2011, the production capacity of PV increased worldwide by 54 percent to meet the growing demand.[2]

When something grows by a percentage each year, it is growing exponentially. Compound interest on a savings account with a fixed interest rate is frequently cited as an example of exponential growth. If the interest earned in one year stays in the account, then that interest earns interest during the next year, compounding the growth rate. With enough time, the accumulated interest can overtake and exceed the original amount.

Exponential growth is much faster than linear growth, because instead of growing by a fixed *amount* each year, it grows by a fixed *percentage* of the previous year. Since each year's total is bigger than the previous year's, the absolute amount added each year is bigger. In other words, the annual growth is growing.

A graph of exponential growth always curves upward over time, simply because each year's growth is larger than the previous year's growth.

When you have exponential growth, it's only a matter of time before the original amount doubles. With a high- percentage growth rate, the *doubling time* is short, while a low-percentage growth rate has a longer doubling time. So doubling time is another way of expressing the rate of exponential growth.

In exponential growth, the doubling time can be estimated by dividing the number 70 by the percentage growth rate.[4] For example, if a business grows by 10 percent each year, dividing 70 by 10 yields seven years as the doubling time. So for each seven years, the business doubles in size. A 20 percent growth rate would have a doubling time of only three and a half years.

Doubling is a powerful force if it's repeated enough. You can personally experience fantastic growth through doubling when you make bread. When you knead dough, you flatten it, fold it over, and flatten it again. Each fold doubles the number of layers, and flattening it allows you to keep folding without the dough getting too thick. Instead, each layer gets thinner. After ten folds, you have over a thousand layers (1,024 to be precise). After twenty folds, you have about a thousand thousand, or a million layers. After thirty folds, you have about a thousand million, or a billion layers, so simply by folding bread dough thirty times you can create "Billion-layer Bread."

You see exponential growth and doubling times whenever you have a steady *percentage* growth rate, where each year's gain is proportional to the total from the previous year. Exponential growth is characterized by a small effect in the early stages, and a rapid increase later due to the accumulated doubling. The 11th fold adds 1,024 layers, but the 21st fold adds over a million layers.

Solar PV Exponential Growth

Solar PV is ripe for exponential growth because of the millions of potential participants who could install it. Solar PV is popping up everywhere, and each installation stimulates more people to investigate and install it themselves. This leads to a proportional growth rate for solar PV that will take it from its current level of insignificance to the ultimate goal of displacing all fossil fuels.

The installed base of solar PV in 2012 was 100 GW, and our target is 100,000 GW. Ten doublings would multiply the original by 1,024, pushing the total to 102,400 GW, a bit over our target.

How long would ten doublings take? That depends on the growth rate. If the annual growth rate of solar PV was 20 percent, then the doubling time would be three and a half years. Ten such doublings would take thirty-five years, so we would arrive at our goal in about 2047.

In other words, if we can maintain a 20 percent growth rate for solar PV, we would have enough PV by 2047 to entirely replace our fossil-fuel consumption, putting the brakes on climate change and ocean acidification.

Such growth rates have been achieved for other commodities, so they are not unprecedented. For example, the number of mobile phone subscribers in the United States grew from 340,000 in 1985 to 300 million in 2010,[5] an average growth rate of 31 percent per year.

Can we achieve anything like 20 percent exponential growth for solar PV?

The amazing good news is that *we already have.* The worldwide PV industry has grown phenomenally for the last decade. Between 2002 and 2012, the installed base of solar PV grew at an average rate of 46 percent per year, faster than mobile phones.[6] The high PV growth rates persisted even through a major worldwide recession, when other investments were being deferred. If the installed PV base continues to grow at a 46 percent annual rate, we would reach 100 TW of PV in only eighteen years, in 2030.

But such high growth rates are probably not sustainable for long periods of time. Products that experience rapid growth typically go through stages of growth, where high initial rates decline as systems scale up. No one can predict such stages, but the following figure outlines one reasonable growth projection, with declining growth rates over several decades:

In this more realistic scenario, we reach 100 TW of PV capacity in 2060. If we manage to stick to a timetable like this, we can achieve 100 percent solar through exponential growth. If mobile phones can go from obscurity to "everyone-has-one" in just twenty-five years through exponential growth, then so can solar PV.

Starting year	Cumulative Installed PV (GW)	10-year average annual growth rate	Cumulative Installed PV after 10 years (GW)
2010	40	35%	804
2020	804	20%	4,980
2030	4,980	15%	20,100
2040	20,100	10%	52,300
2050	52,300	7%	103,000
2060	103,000		

Figure 44. Hypothetical Future Growth of Solar PV

Each line of this table shows the growth for one decade. The end result of each decade is then used as the starting point for the next decade. For example, if the PV growth rate was 35% per year from 2010 to 2020, then the cumulative installed PV would grow from 40 to 804 GW. Starting at 804 in 2020, a 20% growth rate for the next decade would result in 4,980 GW installed by 2030. After five decades, the total would exceed 100,000 GW (which equals 100 TW).

Solar PV Energy Plan

A basic, if odd-sounding, solar PV energy plan for the future might consist of the following:

- Maintain exponential growth in PV production and installations.
- Develop and deploy energy storage.
- Have patience.

The last point is necessary because solar PV is starting with such a small fraction, it will take several decades to reach the goal of 100 percent replacement of fossil fuels. It is the nature of exponential growth that it seems slow in the early years. After two decades, we will have achieved only 5 percent of the goal, yet continuing with exponential growth will take us to the goal in fifty years—only two generations.

Energy storage will become vitally important in the later years of this plan. As more solar PV comes on line, utility companies will need to adjust to accommodate its intermittent nature. Utility companies can manage up to 30 percent of their power coming from intermittent sources, using existing methods.[7] As solar takes over from the last remaining fossil-fuel power plants, energy storage will guarantee reliable power for everyone, just as the batteries do on my home PV system.

"Exponential growth will take us to the goal in fifty years—only two generations"

The most hopeful part of this plan is that we're already carrying it out. We don't have to create a PV industry from scratch as we're trying to do with clean coal. We already have a thriving PV industry, and PV is already growing at phenomenal rates. If we can sustain such high growth rates for several decades, then reaching our goal will become a mathematical certainty.

Sustained exponential growth of PV will be an enormous undertaking. This transition will be building up a massive new energy industry comparable in size to the fossil fuel industry. To maintain these growth rates, new PV factories will need to be built and supply bottlenecks will need to be resolved.

As part of the shift to solar, the factories themselves will gradually transition to run on PV power. Solar breeder factories would allow a solar economy to power its own growth, says solar economist Richard Perez.[8]

Some PV technologies could experience materials shortages as they grow exponentially. The basic silicon material used in most PV cells is the second most abundant element in the Earth's crust, and will never be in short supply. But certain rare earth elements such as indium and tellurium used in thin-film PV are much less abundant. However, a recent report from the UK Energy Research Council concluded:

... there is no immediate cause for concern about availability of either indium or tellurium. There is evidence of considerable potential to increase the extraction of both metals because a sizable proportion of the material potentially available from primary metal extraction is not currently utilized. Moreover, there is potential to increase recycling of products containing indium or tellurium, for example from flat screens. However, the scale of the rollout of PV envisaged in some scenarios could imply a large expansion in the demand for indium and tellurium. There is no reason to believe that this is not feasible, however adequate data on reserves and resources do not exist.[9]

If this 100 TW plan is followed, then we will arrive at a full solar future, but no one can accurately predict events fifty years into the future. Too much can change, especially with new developments in technology. We will mostly be making it up as we go along. The incremental nature of solar PV gives the solar industry the flexibility to learn and improve as it goes. We can follow many leads in solar and energy storage, and continue with those that show the most promise. So only the basic outline of the plan is visible today, with details to be filled in later.

But we must have a *goal*, so we can measure our future decisions relative to our goal. A 100 percent solar-based economy is a worthy goal for human civilization. It may turn out to not be all PV, but it will succeed if it is solar powered.

PART IV: A PLAN

CHAPTER 20

Electrification

Maintaining exponential growth for solar PV will develop the supply side of a solar-based energy system for human civilization. But to complete the transition from fossil fuels, the demand side will have to change as well.

The Institute for Sustainable Development and International Relations summarized this idea as three pillars for deeply cutting carbon emissions:[1]

1. **Energy efficiency.**
2. **Solar electricity**, such as solar PV and wind.
3. **Fuel switching**, that is, switching from fossil fuels to solar electricity or solar fuels.

The biggest change we can expect to see in the transition to solar will be the wholesale replacement of heat engines with electric motors. Today, heat engines in the form of steam turbines generate most of our electricity, and heat engines in the form of internal combustion engines and jet engines power almost all of our transportation. Fossil fuels energize all of these heat engines, making them the largest consumer of fossil fuels. In the United States, for example, 93 percent of all coal consumed is burned in electric-power heat engines, 71 percent of all oil is burned in transportation heat engines, and 38 percent of all natural gas is burned in electric-power heat engines.[2]

Heat engines have proven their value over the last 200 years, but their time is passing. Heat engines have two big problems: They burn fossil fuels, and they are terribly inefficient. The burning of fossil fuels in heat engines contributes the largest portion of human-generated carbon dioxide emissions, making heat engines directly responsible for climate change and ocean acidification.

The poor efficiency of heat engines (25–35 percent) means we must burn three or four times more fossil-fuel energy than the energy we actually use. So two-thirds to three-fourths of the carbon emissions from a heat engine simply pollute without contributing anything useful in balance.

We tolerate such poor efficiency only because heat engines work. They enable us to convert a raw energy form we have in abundance (fossil fuels) into energy forms we find highly useful—electricity and motive power. In the 200 years since the first steam engines were built, no one has invented a fossil-fuel technology better than heat engines for making electricity and motive power.

Retaining all these fossil-fueled heat engines will prevent us from reaching a goal of 100 percent solar-based energy. We cannot fix our energy problem unless we replace the heat engines. Any plan to get off fossil fuels that does not replace heat engines is doomed to failure.

Heat engines are so ubiquitous in modern society that it seems radical to get rid of them. If we simply turned off all heat engines today, then modern society would collapse. So we can't just turn them off, we must gradually replace them. But replacing heat engines is not as hard as you might think.

The heat engines used to generate electricity can be the first to go. My electricity doesn't come from a heat engine. The PV panels on my roof generate electricity directly, without burning fossil fuels in a heat engine. Every PV and wind system that ties into the utility grid provides electricity of equivalent quality to replace the electricity that would have been generated in a fossil-fueled power plant. When combined with energy storage, solar electricity can eventually substitute for all fossil fuel electricity.

What About Biofuels?

Can't we just substitute biofuels for fossil fuels and continue to use heat engines? Not if we want to continue to eat. Wholesale food prices have doubled since 2004. These increases reflect a shortfall in supply to meet demand, but crop production has not slowed, and in fact has grown since 2004. The problem is rapidly rising demand. "Since 2004 biofuels from crops have almost doubled the rate of growth in global demand for grain and sugar and pushed up the yearly growth in demand for vegetable oil by around 40 percent," says Timothy Searchinger of Princeton's Woodrow Wilson School of Public and International Affairs. He adds: "Our primary obligation is to feed the hungry. Biofuels are undermining our ability to do so. Governments can stop the recurring pattern of food crises by backing off their demands for ever more biofuels."[3] Even if we switch to non-food biofuels, the numbers don't work out. The efficiency with which photosynthesis converts sunlight to chemical energy is 1 percent or less, far less than PV panels that produce electricity directly. Then subtract the substantial energy needed to grow the biofuel crop and process it into a useful fuel. Then burn it in a heat engine that is only 25–35 percent efficient. According to George Olah, director of the Loker Hydrocarbon Research Institute, a major part of all agricultural lands would have to be dedicated to energy crops to meet a significant portion of our energy needs.[4]

Heat engines also power most transportation today. Every car and truck that uses an internal combustion heat engine burns fossil fuels to move the vehicle. An electric motor can easily substitute for the engine to create motive power.

Today, you can buy a modern electric car from any of several manufacturers. Despite the limited driving range and relatively long recharge times, people are motivated to buy an electric car because it's cleaner and cheaper to run than a fossil-fueled car.

However, if you charge your electric car from your local utility, it's highly likely that your electric car is still running on a heat engine powered by fossil fuels. Instead of a heat engine in your car, the heat engine is at the power plant, and its energy is delivered via electricity. Instead of using an inefficient heat engine in your car, you're using a slightly less inefficient heat engine at the power plant (typically 33 percent efficient instead of 25 percent for cars).

If you drive an electric car and instead plug it into solar PV panels as I do, then no heat engine is involved, no fossil fuels are consumed, and no carbon dioxide is emitted. An electrified vehicle *enables* solar-powered driving, but does not guarantee it. Until utility companies transition from fossil fuels to solar energy, you'll need to generate your own solar electricity if you want a solar-powered car.

The current generation of electric cars, when charged with solar electricity, will suffice for most local personal transportation. One study of US driving patterns showed that 95 percent of all car trips were under 30 miles, well within the range of existing electric cars.[5]

Those needing a longer range would more likely buy a plug-in hybrid car. Plug-in hybrids, however, still use a heat engine when the batteries get low. As described in the section titled "Liquid Fuel Cells" (page 170), the heat-engine component could be replaced with a fuel cell. If fueled with a solar-produced liquid fuel such as methanol, then it would be free of fossil fuels. Such cars await the further development of methanol fuel cells, as well as a network of filling stations that have pumps with solar-produced methanol.

Trucks could follow a similar pattern. Trucks used only for local deliveries could be 100 percent electric, returning to the base station for recharge. For example, in 2011 the UPS delivery company purchased a hundred electric delivery trucks for their short-range routes in California.[6] Similarly, the US Postal Service is looking to adopt electric delivery vehicles.[7] Since electricity is cheaper than gasoline per mile driven, these companies will save on fuel costs, one of their biggest expenses.

Trucks used for long-distance hauling would need to use a liquid solar fuel cell to provide longer range and quick fillups. Some truck transport may transition to rail transport. In terms of energy use per ton of transported goods, trains are far more efficient than trucks. Steel wheels lose less energy than rubber tires, and air friction is reduced because each train car shields the cars behind it. Existing locomotives could replace the diesel generators with fuel cells to make the electricity to drive their existing electric motors, or, as with many trains, the track itself could be electrified so the train wouldn't have to carry fuel.

Electric motors are even starting to fly. European plane manufacturer Airbus has built a two-seater electric-powered airplane called the E-Fan. Its propellor is driven by a powerful electric motor, with the electricity coming from onboard lithium batteries that can fly the plane for thirty minutes. Airbus plans to develop electric regional aircraft for commercial service in the near future.[8]

Long-distance jet aircraft are unlikely to be electrified in the near future; however, jet fuel can be easily derived from methanol, and methanol can be generated from solar electricity.

High-speed trains provide an alternative to jet travel, and high speed trains can run on solar electricity. One Japanese train puts four electric-wheel motors on each of the train's sixteen cars. The motors are computer coordinated to provide smooth acceleration and deceleration, and make regenerative braking possible.

In some ways, switching from a heat engine to an electric motor is like switching your lighting from incandescent bulbs to LED lights. Heat engines are a 19th-century technology, as are incandescent bulbs. More importantly, just as LED lights are three to four times as efficient as incandescent bulbs, so are electric motors three to four times as efficient as heat engines. The best electric motors are 97 percent efficient. Over time, both heat engines and incandescent bulbs will lose out to more modern and efficient technologies.

Every heat engine that's replaced with solar electricity contributes to the transition from fossil fuels to solar energy. Each increment may seem small, but each replacement permanently shuts down a source of carbon dioxide and air pollution, avoiding many years of pollution.

Electrification will play a key role in controlling climate change. One study showed that deep cuts in greenhouse gas emissions cannot be achieved without widespread electrification.[9]

The transition from heat engines to electric motors isn't new. It continues the overall electrification of human energy systems that has proceeded since about 1900. The manufacturing sector is a good example.

In the 1800s, most factories had a single centralized steam engine to provide mechanical energy. With the introduction of electricity service in the late 1800s, individual machines could each have their own electric motor. In 1905, fewer than 10 percent had been converted to individual electric motors, but by 1930, fully 80 percent had been converted.[10]

That trend continues still. In the United States in 1950, only about 4 percent of all energy used onsite (not for transportation) was delivered in the form of electricity. By 2010, that portion had risen to 29 percent,[11] indicating that electricity is shouldering a greater share of energy tasks.

This gradual adoption of electricity is driven by its many advantages:

- Electricity can be produced from a wide variety of primary energy sources.
- Electricity is easily and efficiently transported, and then easily distributed to individual applications at its destination.
- Electricity can be easily converted to motion, heat, and light.
- Electricity conversions are clean and quiet, with no exhaust fumes.
- Electricity enables new capabilities.

Modern communication serves as an example of the last point. Electricity is the sole energizer of telephones, computers, the Internet, and all electronic entertainment systems.

To reach the goal of eliminating fossil fuels completely, the remaining energy uses beyond transportation and electricity production must be addressed. Since it's unlikely that biofuels can grow sufficiently to replace fossil fuels, electrification will enable those applications to be transitioned.

In general, for each fossil fuel-appliance there exists an equivalent electrical version. Each application of fossil fuels needs to be evaluated and electrified where possible. New installations and the replacement of old or broken equipment provide the best opportunities to make the switch. Over time, all fossil-fueled equipment can be phased out.

There is no doubt that electrification will increase electricity demand. To meet our goal of a 100 percent solar-based economy, that increased demand must be met with solar electricity. If we build more fossil-fueled power plants to meet that extra demand, then we will have accomplished nothing.

Electrification will give us the means to fully phase out fossil fuels, while remaining compatible with our existing energy systems during the transition. An electric car can run on either fossil-fuel electricity or solar electricity, but a gasoline car can run only on fossil fuels. Electrification of all fossil-fueled tasks will make it possible to run them all on solar energy.

CHAPTER 21

Transition Gradually and Gracefully

One of the major strengths of solar PV is that the transition can be done incrementally, on a case-by-case basis. This will allow the transition to proceed smoothly without disruption.

If we instead decided our energy future should be based on nuclear power or clean coal, such plants are feasible only in large sizes. Each giant plant requires years of planning, license review, and construction before it can generate a single kilowatt-hour of energy.

Rather than lurching in big steps to a new energy system, solar PV allows a gradual shift, in millions of tiny steps. "Small, quickly built units are faster to deploy for a given total effect than a few big, slowly built units. Widely accessible choices that sell like cellphones and PCs can add up to more, sooner, than ponderous plants that get built like cathedrals," says energy expert Amory Lovins.[1]

We can co-power our economy with fossil fuels and solar during the transition. Initially we'll use mostly fossil fuels, but over time, as more solar PV comes on-line, the solar fraction will grow and the fossil-fuel fraction will shrink.

A transition to solar-based energy carries less risk because it's carried out in much smaller steps, allowing for course corrections as we proceed. At each stage of the transition, it's likely that new problems will arise. The modular nature of solar PV and the short lead-time to install it make it easier to correct for those problems as they come up. For example, if a certain new type of PV panel starts failing after a few years, the manufacturing problem can be corrected and replacement panels can be installed quickly.

Contrast that with fixing a newly discovered nuclear safety problem. A new design for nuclear power plants takes years to develop and test, and we wouldn't know how well it works until several are built.

The difference is in how long it takes for new learning to improve the system. With solar, that's a year or two; with nuclear, the time frame is decades; with clean coal, no one knows because large-scale plants are just now beginning construction.

The exponential growth pattern of solar PV helps too. Exponential growth starts slow, allowing us to resolve any problems when volumes are low, before we become too dependent on them. And the gradual rollout allows for incremental planning for each increment of change.

By gradually phasing in solar PV and resolving problems as we go, the transition from fossil fuels to solar will not be disruptive to the economy. Governments fear that cutting carbon dioxide will slow their economic growth, so they don't sign onto international treaties that commit them to such cuts. Certainly, fossil fuels do not have a future growth path, but solar does. Economic growth can continue if it's powered by solar PV.

And we have shifted energy sources before without disruption. Zoom your energy scope back in time, and you can see that in the 1800s we shifted from water and animal power to coal-powered steam engines. In the 1900s we completely shifted from horse-powered transportation to oil-powered transportation. If in the 2000s we shift from fossil fuels to solar energy, we'll be continuing this logical progression.

During each of the previous transitions, both old and new energy sources were in play at the same time. The old source was in decline, and the new source was growing. Eventually the new source grew sufficiently to take over, allowing the old source to be retired.

The current transition from fossil fuels to solar-based energy is in its early stages. Solar PV barely shows up on the curve, and fossil-fuel use is still growing in countries such as China and India. But solar PV is growing at an exponential rate, and its effects will soon be felt.

The Pursuit of Permanent Power

CHAPTER 22

Shift to Power Sources

The transition from fossil fuels to solar-based energy differs from all previous energy transitions in a deep and fundamental way. We are replacing energy stores with permanent power sources. The implications extend far into the future.

The Earth holds stores of coal, oil, and natural gas, provided by biology and geology. We, the energy-using animal, have been drawing from those energy stores at an ever-increasing rate, allowing us to build and run our modern civilization. But it takes millions of years to convert plant and animal matter to fossil fuels, so they're not being replaced at a rate that's useful to us. Our initial store of fossil fuels is all we get, and we've been drawing down that store for about 200 years.

Solar energy is *not* an energy store. Solar energy is continuously generated by the fusion reactions in the core of the sun. The sun's energy flows constantly to Earth, where it warms the planet and is available to be converted to electricity, if we choose to do so.

A continuous flow of energy defines a *power source*. Power measures the *rate* of energy flow, as in joules per second. So a constant energy flow is best described by its rate of flow, and as a source of power rather than a source of energy. If an energy store is like a static lake, then a power source is like a flowing river. Power is energy in motion.

A solar PV panel exemplifies a power source, putting energy in motion when the sun shines. The electricity that flows out of it has to be either used directly or converted to another form for storage.

Several practical features distinguish a power source from an energy store. The principle difference is how long each lasts. A power source maintains a steady flow and doesn't run out. The sun is good for another 5 billion years of power, essentially forever in human terms. In contrast, our fossil fuel energy stores are finite, and, regardless of what rate we them, will eventually be depleted. If we did manage to use fossil fuels for another couple of centuries, then we would completely use up the stores.

The world economy runs twenty-four hours per day, seven days per week, year after year, continuously flowing large amounts of energy to carry out all those economic activities. Any energy need that's continuous is best matched by a power source—one that supplies energy continuously. When we try to meet a continuous energy need with energy stores as we do today, we use up those stores and must continuously seek new ones. The world is littered with drained oil wells and abandoned coal mines as energy companies move on to fresh sources. If production falters for any reason, then supply cannot meet demand, anxieties rise, and energy prices skyrocket.

Compare that to a world economy running on solar energy. The continuous flow of energy needed for economic activities would be supplied by a continuous flow of energy from the sun. A power source like the sun is better matched to a continuous world economy. As long as the sun rises, the economy flows.

An energy store and a power source also differ in *how fast* you can use them. You can pull energy out of an energy store as fast as you want.

> *"As long as the sun rises, the economy flows."*

Of course, the faster you use the energy, the faster the supply diminishes, but that's a choice you have with an energy store. In contrast, a power source establishes a fixed rate of energy flow, and you can't use energy faster than the power source provides it.

So with an energy store like fossil fuels, you can burn energy as fast as you want, until you burn it all up. With a power source like the sun, you can use energy only as fast as it comes in, but it comes in forever. You can liken it to two siblings spending their inheritance. One wildly spends their windfall inheritance on a lavish lifestyle until the money runs out, while the other wisely invests their inheritance and lives more modestly off the investment income, which does not run out. Solar energy is like that steady income. If you can live on your income, you can live forever.

The first head of the US Environmental Protection Agency, William Ruckelshaus, summed it up neatly in 1990: "Nature provides a free lunch, but only if we control our appetites."

A power source also has a unique characteristic not present in any energy store. With a power source, you must *use it or lose it*. If the sun shines and you don't use or store its energy, then you lose it. With a power source like sunlight, energy is already flowing. If you want to use it, you must actively divert some of that flow to serve your purpose. If you don't, then the energy will flow anyway, but it won't serve your purpose. This rule does not apply to energy stores, because they're not lost if they're not used. They're saved for a later day.

I occasionally experience this use-it-or-lose-it rule in the operation of my own off-grid solar PV system. When my house and car batteries are full and I have no other need for electricity on a sunny day, then I'm losing energy. Since I'm not connected to the grid, my solar electricity has nowhere to go, so the controlling electronics shut off the flow of electricity from the panels. The solar energy falling on my PV panels is no longer being transformed into electricity, so it's just converted to heat on the panel surface. At that point, my PV panels are acting like big roof shingles, albeit very expensive ones.

The amazing feature of solar energy, however, is that my "loss" is of no great consequence. When a gasoline-powered car idles at a stoplight, it still burns gasoline to keep the engine running. That wasted gasoline performs no useful work, yet it still emits air pollution and carbon dioxide, and it still contributes to climate change and ocean acidification. My wasted solar energy does not have any such effect on the environment.

That's because I'm not wasting energy as much as I'm wasting an opportunity. The solar energy flows out of the sky and onto my panels whether or not I use any electricity that comes out of them. When I use the electricity, the energy flows through my appliance, serves some purpose, and is eventually degraded to tepid heat through friction. If I don't use it, electricity does not flow, and the solar energy degrades to tepid heat directly on my PV panel. The end result is the same: solar energy converted to heat. The difference is: do I seize the opportunity or not?

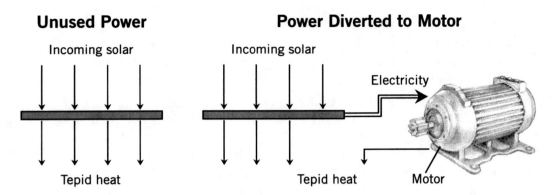

Figure 45. Putting solar power to use
Solar energy flows whether or not it's used. In both cases, the energy ends up as tepid heat, but if properly diverted, the energy can perform a useful function before it gets there.

The solar energy flowing onto the Earth presents an opportunity for the human race to plug into a vast power source. When you set up solar PV or wind turbine equipment, you directly connect to a 384 yottawatt power source, the sun. Your equipment transforms the raw solar power source into an electricity power source that can be easily transmitted and used by anyone.

Energy Conservation vs. Energy Efficiency

A power source also draws a clear distinction between *energy conservation* and *energy efficiency*. When applied to fossil fuels, *energy conservation* means to conserve the store of energy, to use it sparingly, to prolong its life. That meaning doesn't apply to a power source, where the energy flows whether you use it or not, so there's nothing to conserve. With a power source, the term *energy efficiency* makes more sense when used to describe how well you apply the already flowing energy to meet your needs.

Modern society is just learning to use power sources. Most existing energy systems are based on 200 years of experience with fossil-fuel energy stores. Our major energy-production systems today consist of mining or drilling operations to extract the fuel material, transportation systems to move the huge amounts of material, refining systems to cleanse it, storage systems to hold it in readiness, combustion systems to transform the chemical energy into heat, and heat engines to convert the heat into motion or electricity.

With a power source such as solar PV, there are no energy materials to transport, store, or burn. Rather, once the PV panels are installed, there's not much to do except *use* the energy as it comes out of them. That sounds easy enough, but it takes some adapting to make it work.

PV panels make electricity when the sun shines, not necessarily when you need to use energy. If you happen to need energy when the sun shines, then that's the optimal use of such a power source, as the energy flows directly from source to end use. But if your energy needs are not synchronized with the sunshine, as many are not, then some form of energy storage is needed, as described in chapter 13, *Energy Storage*. A power source combined with energy storage provides the best of both worlds—energy on demand, like fossil fuels, and energy that lasts forever, like solar.

Because a power source is not depleted as it's used, it encourages its own use. Since the energy flows anyway, you might as well put it to some good purpose. Not using a power flow is the only way to waste it. And installing more PV panels to grow the power source does not deplete it either.

Contrast that with fossil fuels, whose use we must discourage to preserve the energy stores for future generations and to avoid degrading the environment. There's no guilt associated with using a solar power source, and encouraging it to grow will help the environment.

"Not using a power flow is the only way to waste it."

Since we must base our economy on energy, wouldn't it be better to base it on a type of energy that you want to encourage to grow? If we want a growing economy, it makes more sense to base it on a power source whose growth has positive effects.

Solar transforms the act of using energy from a negative to a positive. Fossil fuels have so many bad consequences that people get the idea that energy use in general is a bad thing that must be curtailed. But it's not energy use per se that must be minimized, only fossil-fuel energy use. We can use nearly unlimited amounts of solar energy without negative effects.

With solar energy, there's no need to hold back on energy use. A permanent power source provides an abundant and unending flow of energy. At the same time, it makes no sense to waste the solar energy, since you've invested in the equipment to transform it for human use.

I consider the solar electricity I generate to be precious, not only because I produce it myself, but also because solar's many good qualities give it a higher value than the equivalent fossil fuel energy. So I use only efficient lights and appliances to get the most from this precious resource.

I describe my solar electricity as both abundant and precious. In economic terms, those two words, abundant and precious, would be contradictory. Generally any commodity that is abundant is cheap, not precious. But solar's abundance comes over time, not all at once, since solar equipment can generate power indefinitely. And its preciousness exists mostly outside of the conventional economic accounting system, as I described earlier in chapter 16, *The Compelling Economics of Solar PV*.

The economic equations make more sense when you acknowledge that solar energy is not a commodity. There's no store of solar energy that can be mined, metered, and sold to the customer. Raw solar energy is delivered everywhere, and a "customer" can simply make their own electricity from it. The futures market for solar energy is on your roof, not in the stock market.

These fundamental differences between fossil fuels and solar energy are obscured by the terminology we use to describe them. When we describe solar energy as *renewable energy*, we mean the energy source is renewed rather than depleted. This terminology describes solar in terms of energy stores, as if solar were another kind of fossil fuel that magically regenerates itself.

But solar PV and wind do not come to us as energy stores, they come as power sources. There is no energy store to be depleted or renewed, because there is no energy store at all. The energy actively flows from the sun, and we have the opportunity to divert some of it for our own purposes. Nothing is depleted if we use it, and nothing is depleted if we don't use it. Solar power is nothing like our fossil-fuel energy stores.

Calling solar power "renewable energy" is like calling an automobile a "horseless carriage." While it relates the new to the familiar, it frames the new in terms of the old, and so fails to describe the new on its own terms.

We can describe solar on it own terms as *permanent power*. That term distinguishes solar as a power source, and captures the fact that it's permanent, its most innovative and important feature.

Of course, a piece of solar PV hardware is not permanent. It may last twenty to thirty years, but it will need to be replaced periodically. What can be permanent, though, is a physical area staked out for solar energy. This could be an area of rooftop, highway median, or desert land. Once PV equipment is installed, it establishes a claim to that solar space, a claim that can be maintained by updating of the solar hardware as needed. With such updates, that space can generate power forever.

> *"We can describe solar on it own terms as* permanent power*"*

Despite its inadequacies, the term "renewable energy" is well entrenched and unlikely to be replaced. But when you read or hear the term "renewable energy," mentally transform it to "permanent power," and it will give you a new perspective on the great potential for solar-based power systems.

PART V: THE PURSUIT OF PERMANENT POWER

CHAPTER 23

The Solar Epoch Begins

This shift from fossil fuels to solar power sources represents more than a change of energy source, it is a change of method. We're moving from an energy system to a power system. In our existing energy system, we draw from energy stores at unlimited rates. With solar, we engage permanent power sources that deliver energy at fixed rates, but do so indefinitely.

Why is that significant? Once we establish permanent power, we won't need to go looking again in fifty or a hundred or a thousand years for a new source of energy. This will be the last energy transition that we need to undertake.

Transitioning from finite energy stores to permanent power sources represents a fundamental change in the human relationship with energy. This change looms larger than the discovery of fire, the development of agriculture, and the exploitation of fossil fuels.

Our identity is shifting from "the energy-using animal" to "the power-using animal." This unprecedented shift marks a distinct turning point in history, the start of a new energy epoch.

Our first energy epoch, the Wood Energy Epoch, began when prehumans learned to control fire about a half million years ago. It lasted for five-thousand centuries, up through the early years of the Industrial

Revolution. As the energy-using machines of the Industrial Revolution multiplied, the energy sources of the Wood Energy Epoch could not keep up. Those limitations were overcome with the introduction of fossil fuels and heat engines, marking the beginning of the second energy epoch, the Fossil Fuel Epoch.

The growth in the use of fossil fuels gradually led to the growth of the problems associated with them. For most of the Fossil Fuel Epoch, those problems were tolerated in order to get the energy. Only now, as we witness significant global changes wrought by carbon dioxide emissions, are limits being proposed for fossil fuels. Climate change and ocean acidification are forcing us to reexamine fossil fuels and search for alternatives. Now it must be only a matter of time before we phase out fossil fuels and draw an end to the Fossil Fuel Epoch.

The rise of a third energy epoch will enable the Fossil Fuel Epoch to decline. The third epoch will be based on the direct use of the sun as a power source. It can only be called the *Solar Epoch*.

The Solar Epoch will employ a variety of solar energy forms, such as solar PV, solar heat, wind, hydropower, waves, and sustainable biofuels. The future contribution of each of these technologies is yet to be determined, given the rapid development taking place on many fronts. From today's perspective, solar PV will most likely predominate, with the other solar power sources acting as supplements.

The Solar Epoch is distinctly defined by the shift from finite energy stores to permanent power sources. Every PV and wind installation moves us an increment into the Solar Epoch.

As more solar-based power sources are brought online, the rate of withdrawals from fossil-fuel energy stores will decline. At the end of the transition period, the world will be operating on continuous power sources, and what remains of the fossil fuels can stay in the ground.

Each of these energy epochs came about gradually, so it is not possible to assign a specific starting or ending date to any of them. The beginnings of the Wood Energy Epoch are lost in time, and in some parts of the world, wood energy is still used. Rather, the Wood Energy Epoch declined gradually from its peak, while fossil fuels grew. As the Fossil Fuel Epoch's influence expanded, that of the Wood Energy Epoch contracted. These two epochs overlapped during the transition period.

Likewise, the Solar Epoch will overlap the Fossil Fuel Epoch during its transition. Any expectations that we can suddenly shut off fossil fuels and start using only solar-based energy are not realistic. The process can and must be gradual so as to not disrupt the current economic and political systems. Instead of cutting back on activities that require energy, we substitute replacement energy without stopping any economic activity. History has shown us that such a gradual energy transition is possible.

When will the Solar Epoch begin? Actually, it already has. We are in the early stages of the Solar Epoch now. Each installation of solar equipment becomes a power source that displaces energy stores. At the present time, solar is contributing a small percentage of our energy needs, so it's hard to notice, but the exponential growth curve has started.

The cosmic significance of the Solar Epoch is that it need not end. By shifting from energy stores to permanent power sources, there will be no need for a fourth energy epoch. The Solar Epoch can be sustained for the lifetime of our star.

By comparison, the Fossil Fuel Epoch will be seen by future historians as the shortest of the three energy epochs. The Fossil Fuel Epoch began to have significant influence around 1800, and will largely end by 2100 if we are to avoid catastrophic climate change. The Fossil Fuel Epoch's total span of 300 years looks brief compared to the 500,000 years of the Wood Energy Epoch that preceded it. The Solar Epoch could last much longer than that, as long as five billion years.

"The Solar Epoch can be sustained for the lifetime of our star."

If you measure these time spans in human generations (assuming twenty-five years per generation), then the Wood Energy Epoch spanned roughly 20,000 generations, the Fossil Fuel Epoch only twelve generations, and the Solar Epoch could, in theory, last 200 million generations.

Taking the long view, the twelve generations of the Fossil Fuel Epoch will be seen as a short blip in the human relationship with energy. It's an important blip, though, because each energy epoch enabled its successor to develop. The Wood Energy Epoch enabled the Fossil Fuel Epoch to begin. Without the metal tools developed during the Wood Energy Epoch, the machinery for large-scale coal mining or oil drilling would not have been possible.

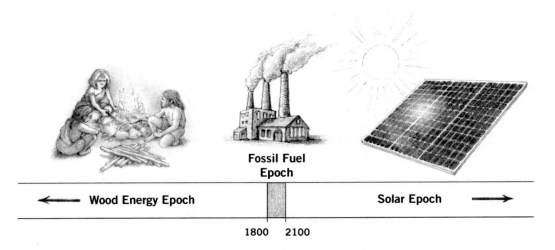

Figure 46. Energy epoch timeline
The Wood Energy Epoch lasted about 500,000 years, while the Fossil Fuel Epoch will last only about 300 years. The Solar Epoch could last for billions of years.

Similarly, the Fossil Fuel Epoch has enabled the Solar Epoch to begin. Fossil fuels drove the Industrial Revolution that made our current high technology possible. That technology has led to efficient wind turbines and solar PV panels, the initial tools of our next energy epoch.

In that long view, fossil fuels become just an interim energy source, a set of energy training wheels for the human race to apprentice with. Fossil fuels are clearly not sustainable in the long run, though fossil fuels can play a role in boosting us up to an energy system that is sustainable.

For all the reasons listed in the section titled "The Many Positives of Solar PV" (page 186), solar is the best energy source for the human race. Fossil fuels should be viewed as our steppingstone for getting to that solar-based energy system. If we use up all our fossil fuels without getting to solar, then we will have wasted that opportunity, and we may not get another chance.

The world will be enriched, not impoverished, with the advent of the Solar Epoch. Here are some of the ways the world will be improved:

- The billion or so people currently without electricity will have it.
- Cities will have clean air and blue skies.
- Ocean waters will be free of oil spills and acidification.

- The main driving force behind global warming will be tamped down.
- Economic growth can proceed without being hampered by carbon restrictions.
- National economies will be more stable because most of the energy they depend on will be generated within their borders.
- Conflicts over oil in the Middle East can be wound down.
- Nuclear proliferation can be brought under control.
- Individuals, businesses, government entities, and organizations will generate income from their solar arrays.

These improvements are huge because energy is huge in our society. Changing the fundamental driving force behind civilization gives us the opportunity to fix formerly intractable problems. When we shift off of the dirty energy sources we currently use, the problems they cause will fall away. We will not be restricting carbon emissions, we will be abandoning them.

"We will not be restricting carbon emissions, we will be abandoning them."

Solar-based energy systems will make us a truly advanced civilization. We will eventually see that fossil fuels are primitive, dirty, and dangerous. Abandoning fossil fuels in favor of solar is a step forward for human kind. Wouldn't a truly advanced society choose an energy system that is clean and lasts forever, given the choice?

Yet the transition to the Solar Epoch is not a sure bet. While solar PV is growing exponentially, the burning of fossil fuels continues to grow as new methods of extraction such as fracking sustain the supply. Which trend will win in the long run is yet to be determined.

PART V: THE PURSUIT OF PERMANENT POWER

CHAPTER 24

What Holds Us Back?

ost people do not choose their energy source. Modern energy systems are based on the consumer model—large energy corporations produce energy as a commodity for sale, and customers buy energy when they need it. Traditional energy corporations fall into two groups: fossil fuel producers like oil, gas, and coal companies, and electricity producers in the form of utility companies that provide electricity service to a wide area. Those types of companies make available most of the energy used in modern society.

In this model, consumers have little to do with the choice of energy sources. They use electricity, natural gas, and gasoline as it is supplied to them, without any control over the sources of the energy. Consumers rely on energy corporations to make energy available, so if consumers want solar energy, they must wait for the energy corporations to adopt it as an energy source. How likely is that?

Fossil-fuel corporations certainly could be a major force in a transition to solar energy because they control vast resources worldwide.

But corporations built for fossil fuels are unlikely to embrace solar PV. Solar PV is intended to replace fossil fuels if we are to make progress on controlling climate change and ocean acidification. Thus solar competes directly with their primary products, from which they make bil-

Rank	Corporation	Business
1	Royal Dutch Shell	Oil
2	Wal-Mart Stores	Retail
3	ExxonMobil	Oil
4	Sinopec Group	Oil
5	China National Petroleum	Oil
6	BP	Oil
7	State Grid	Electric utility
8	Toyota Motor	Automobile manufacturer
9	Volkswagen	Automobile manufacturer
10	Total	Oil

Figure 47. 2013 Fortune Global 500 Top Ten[1]

Of the ten largest corporations in the world in 2013, six were oil companies. Two more were automobile manufacturers, which depend on oil. China's State Grid electric utility is also heavily dependent on fossil fuels.

lions in profits. Since the fundamental motivating force of corporations is to maximize profits for their shareholders, shifting to solar would happen only if they could increase profits by doing so.

The current economic climate does not favor such a shift. As described in chapter 16, *The Compelling Economics of Solar PV*, fossil fuels carry with them many hidden costs that fossil-fuel corporations do not have to account for. That makes fossil fuels cheaper than they should be.

And as described in the same chapter, most of the benefits of solar that make a difference in the world are not currently assigned any economic value. Solar's assets, which include arresting climate change, improving national energy security, clearing skies over cities, and keeping water unpolluted, do not show up on any corporate balance sheet, and therefore make no difference in corporate financial decision making.

Thus the profits to be made on fossil fuels will continue to exceed those of solar PV as long as fossil-fuel costs remain hidden and solar benefits remain undervalued. Fossil-fuel corporations have no motivation to change.

Now, a formerly ignored side effect of the fossil-fuel industries, carbon dioxide, is forcing change. Most national governments and all major scientific organizations have acknowledged that climate change and ocean acidification driven by carbon emissions are global problems that need international action.

The report of the UN Intergovernmental Panel on Climate Change[2] in 2013 summarized the current situation:

- Greenhouse gas emissions are surging, not declining.
- Recent global temperatures demonstrate human-induced warming.
- The melting of ice sheets and glaciers is accelerating.
- Arctic sea ice is in rapid decline.
- Current sea-level rise is higher than originally estimated, forcing sea-level predictions upward.
- Delay in action risks triggering positive feedback effects leading to irreversible damage.
- The turning point must come soon.

The mounting evidence is forcing national governments to consider policies to mitigate climate change. Government intervention could take the form of assigning economic costs to carbon emissions. That would shift at least one hidden cost from the shadows into the light. Yet many energy corporations actively resist such efforts. ExxonMobil spent $16 million funding front organizations to sow doubt about the science of climate change,[3] and oil tycoons Charles and David Koch fund groups that promote the denial of climate change.[4]

Most oil companies also maintain large lobbying efforts to try to thwart legislation they judge to be against their interests. In the United States in 2010, oil and gas corporations spent $146 million lobbying Washington D.C.[5] Over the years, these lobbying efforts have opened up offshore oil drilling, stopped oil tax increases, and killed efforts to reduce government subsidies to oil corporations. And more recently they began trying to halt the spread of solar-based energy, indirectly acknowledging that solar has grown large enough to be a threat to their business interests.

Motivated by popular concern about climate change, national leaders periodically announce that they are committed to reducing carbon dioxide output, yet they have repeatedly backed away when serious carbon restrictions are put to paper. Why? Their reluctance comes from a chain of logic that goes something like this:

a.   Our economy is entirely dependent on fossil fuel energy.

b.   Limiting fossil fuel use will therefore harm the economy.

c.   I can't be blamed for harming the economy.

Roger Pielke Jr., a political scientist at the University of Colorado, calls it the iron law of climate policy: When there is a conflict between policies promoting economic growth and policies restricting carbon dioxide, economic growth wins every time.[6]

The result is that national leaders defer taking action. That keeps them on record as opposing climate change, yet relieves them of any potential blame. Doing nothing about climate change carries less political risk than doing something. These attitudes have led to the current paralysis that hinders any progress on halting climate change.

Politicians are particularly sensitive to the charge that taking action will put their country at an economic disadvantage compared to other countries that do not enact such carbon restrictions. The solution would be to enact international standards that all countries adhere to. Since air moves without regard to national boundaries, fossil fuels burned anywhere contribute CO_2 to the atmosphere that all nations share. Controlling climate change should spur cooperation among all nations, which would give cover to politicians enacting legislation within their countries.

Such cooperation began in 1992 at the Rio Earth Summit, when the United Nations Framework Convention on Climate Change (UNFCCC) was created. The UNFCCC is an international treaty whose objective is stabilizing greenhouse gases in the atmosphere. Although signed by most energy using nations, the treaty by itself has no target goals or enforcement mechanisms. Those were to be negotiated later in extensions to the treaty, called *protocols*.

The first protocol was negotiated in Kyoto, Japan, in 1997. The Kyoto Protocol was seen at the time as just a first step, and set target levels only for the forty largest industrialized countries. The reasoning at the time was that these countries were responsible for most past and current emissions of greenhouse gases, and that developing countries would be addressed in a later protocol.

To gain acceptance, the protocol set only modest targets. Overall, the global target for the rate of emission of greenhouses gases in 2012 was to be 5.2 percent below the emission levels in 1990. That is approximately equivalent to the emission levels in 1987. Since emissions in 1987 were already raising the levels of greenhouse gases in the atmosphere, it was clear even then that further reductions would be needed to stabilize those concentrations. If you view the accumulating gases as a time bomb for climate change, then the Kyoto Protocol simply set back the clock on the bomb without defusing it.

Even with such modest goals, national governments hesitated to ratify the protocol. It was not until 2005 that enough countries ratified so it could go into effect. With an expiration date of 2012, the Kyoto Protocol participants had only seven years to meet their targets.

The United States remained the biggest holdout. The US Senate preemptively refused to ratify the Kyoto Protocol. The Byrd-Hagel Resolution passed the Senate five months before the Kyoto Protocol was completed, and said quite explicitly that the refusal was for economic reasons:

> Whereas the Senate strongly believes that the proposals under negotiation, because of the disparity of treatment between Annex I Parties [developed countries including the US] and Developing Countries and the level of required emission reductions, could result in serious harm to the United States economy, including significant job loss, trade disadvantages, increased energy and consumer costs, or any combination thereof.
>
> —Byrd-Hagel Resolution, U.S. Senate Resolution 98, July 25, 1997

Since no politician wants to be blamed for hurting the economy, the Byrd-Hagel Resolution passed the Senate 95–0. Since then, no US president has even submitted the Kyoto Protocol for ratification, not even Bill Clinton who helped negotiate the treaty.

Worldwide, carbon dioxide emissions from energy consumption in 2011 had risen a whopping *49 percent* higher than 1990.[7] And it could have been even worse. During that period, countries formerly in the Soviet Union experienced extreme economic downturns, resulting in greatly reduced energy use and CO_2 emissions in those countries. Those reductions were, however, overwhelmed by massively increased emissions of rapidly developing countries, led by China and India. In 2007, China surpassed the United States to become the largest emitter of carbon dioxide in the world.[8]

Since actual emissions are going in the opposite direction of the intentions of the Kyoto Protocol, the UN recognized that further protocols were needed, in order to include developing countries, to set stronger goals and enforcement mechanisms, and to continue the treaty past 2012 when the Kyoto Protocol expired. A UN-sponsored meeting in Copenhagen Denmark in 2009 was supposed to finalize a successor to the Kyoto Protocol, but nations could not agree on any substantial reductions. A follow-up conference in Cancun Mexico in 2010 had the same lack of results.

Hopes that were attached to the 2011 conference in Durban, South Africa were deflated by its outcome. The conference agreed to essentially start over in negotiating a climate treaty, to be finalized by 2015 and coming into force by 2020. With the Kyoto Protocol expiring in 2012, that leaves eight years with no binding commitments at all. Achim Steiner, the director of the UN Environmental Programme, says that carbon emissions must *peak* in 2020 to prevent irreversible global warming. Yet with this treaty, the cuts will not even *begin* until 2020.

Without an international treaty to back them up, national governments are struggling to take actions on climate change that might be seen as harming their economy or making them less competitive in world markets. To make matters worse, many countries have continued to subsidize fossil fuels.

For decades, governments have encouraged and subsidized fossil fuels to promote economic growth. There are generally two types of fossil fuel subsidies: production subsidies and consumption subsidies. Production subsidies help fossil fuel companies produce energy more cheaply, with the goal of making a nation more self-reliant on energy. For example, US fossil-fuel corporations receive billions in tax reduc-

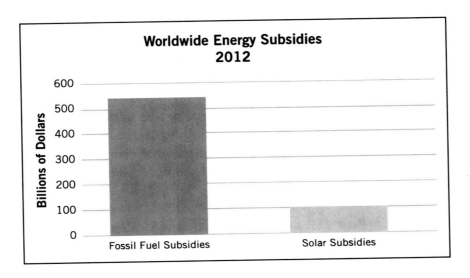

Figure 48. Fossil fuel versus solar subsidies
Worldwide fossil fuel subsidies are five times larger than all solar energy subsidies. Gradually reversing such policies would accelerate the shift from fossil fuels to solar energy.

tions from the Foreign Tax Credit and the Oil and Gas Excess Percentage over Cost Depletion deduction.[9] Production subsidies are used primarily in industrialized countries.

Consumption subsidies keep the retail price of fossil fuels lower than market rates. Consumption subsidies predominate in developing countries, to prevent high energy prices from inhibiting nascent economic growth. For example, Iran's government subsidized the price of oil products by 89 percent, leaving consumers to pay only 11 percent of the market rate in 2009.[10]

Governments feel it is their responsibility to ensure a plentiful supply of "affordable" (cheap) energy to keep their economies humming along. Oil, in particular, is considered critical to a modern economy, so it gets substantial subsidies. The International Energy Agency reported that worldwide, fossil fuels received $544 billion in subsidies in 2012, while solar-based energies received only about $100 billion.[11]

How National Governments Can Help

Since politicians claim that cutting carbon emissions will harm their economies, few nations have actually met their carbon-reduction goals. This impasse can be broken by approaching the problem from a different direction.

When you zoom out your energy scope to take in the bigger picture, it becomes clear that human energy use is driving climate change, which emerges as a side effect of our fossil fuel-powered energy systems. Trying to set limits on the side effect fails to address the source of the problem.

Instead of focusing solely on carbon dioxide, nations should be moving to power their economies with solar energy. There is no question that we need energy to run our economies, but we needn't depend only on fossil fuels, because we have solar-based alternatives. As we cut back on energy use to reduce carbon emissions, we can also substitute solar-based energy over time, thereby maintaining full energy flow to run a vibrant world economy.

Climate change is a compelling reason to move to a solar-based economy, but it's not the only reason, nor need it be the primary reason. National leaders could make more progress if they focused on all the positive benefits of switching to solar-based energy systems. Then fixing climate change could be seen as just a side benefit of investing in solar.

Governments have a range of options to facilitate the move to a solar-based economy, including:

- Removing government subsidies for fossil fuels.
- Compelling electric utility companies to shift from fossil fuels to solar.
- Charging a carbon tax for fossil fuel use.
- Establishing feed-in tariffs for solar-based electricity.

The first step governments should take to facilitate solar investment is to reduce or eliminate subsidies for fossil fuels. "With $55 oil, we don't need incentives to the oil and gas companies to explore. There are plenty of incentives," said President George W. Bush, himself a former oilman, in 2005.[12] Since then, oil prices have hovered around $100 per barrel. Fossil-fuel subsidies distort the energy marketplace, unfairly favoring what are viewed today as the wrong kinds of energy. Remove the subsidies, and solar energy can compete on a level playing field.

Even better: Shift the fossil fuel subsidies to support solar energy instead. A 2013 study by the International Institute for Applied Systems Analysis showed that such a shift would substantially meet the clean energy investment requirements to hold global warming to under 2 degrees Celsius.[13]

Shifting fossil-fuel subsidies would make it easier for electric utility companies to justify investments in solar instead of new fossil fuel plants. Some governments are accelerating that process by mandating that a certain portion of the electricity supply be solar-based by a certain date.

For example, the California Renewables Portfolio Standard requires that in the overall portfolio of electricity sources, 33 percent of electricity in California must come from permanent power sources by 2020. To meet these requirements, some utilities are building their own large solar arrays, and some are buying power from solar or wind installations that are being developed by private interests who can make use of tax incentives. An example of the latter is the 48-megawatt Copper Mountain Solar facility set up near Boulder City, Nevada, in 2010 by Sempra Generation. It contracted to sell the power to California utility company Pacific Gas & Electric.

Such Renewable Portfolio Standards have been enacted in ten countries and fifty other jurisdictions, including thirty US states. British Columbia now requires 93 percent of new power capacity to be from permanent power sources.[14] The European Union's Directive on Electricity Production from Renewable Energy Sources mandates an overall goal of 20 percent solar-based electricity by 2020, but leaves implementation up to individual countries. Germany is pursuing a longer term goal of 80 percent solar electricity by 2050.

Governments can further level the economic playing field by bringing forward the hidden costs of fossil fuels. Since the actual costs are so difficult to determine, governments can simply place a tax on carbon-based fuels to compensate for the harm they do.

A carbon tax would add a small percentage to the price of any energy product based on the amount of CO_2 it releases into the atmosphere. Coal-generated electricity produces the most carbon dioxide per unit of energy delivered, and so it would have the highest carbon tax, while solar PV electricity would have none (although the tax would be paid on any fossil fuels used to make the panels).

A carbon tax is usually designed to start low so as to not suddenly disrupt the economy, and then grow over time according to a published schedule. By publishing the schedule, fossil-fuel users could decide for themselves when would be the best time to reduce energy consumption and install solar equipment. The longer they wait, the harsher the tax becomes.

A carbon tax is easy to calculate, and easy to collect since it applies to energy products that are already subject to various taxes. Its simplicity contrasts with carbon trading schemes, in which allocations for carbon emissions are traded in financial markets. Carbon trading systems require regulators to estimate and track carbon savings for various activities that can cross national boundaries. Some worry that the complexity of the system allows unethical traders to "game" the system for financial gain without producing significant carbon reductions.[15]

The US National Academy of Sciences recommends a carbon tax: "... analyses suggest that the best way to amplify and accelerate such efforts, and to minimize overall costs (for any given national emissions reduction target), is with a comprehensive, nationally uniform, increasing price on carbon dioxide emissions, with a price trajectory sufficient to drive major investments in energy efficiency and low-carbon technologies."[16]

Even energy giant ExxonMobil agrees that a carbon tax would be more effective. Their published climate view states that if a price is set on carbon emissions, then "... it is our judgment that a carbon tax is a preferred course of public policy action versus cap and trade approaches."[17] They suggest the carbon tax be "revenue neutral," which means the money collected under the tax is redistributed back to consumers to soften the impact of the tax, and to make passing a new tax politically feasible. The Citizens' Climate Lobby promotes such a tax.[18] British Columbia adopted a carbon tax in 2008 and after six years, per-person fuel use had dropped by 16 percent (while it rose by 3 percent in the rest of Canada).[19]

A carbon tax could be far more effective if its revenues were expressly diverted to help pay for PV and other solar-based energy. That would essentially double the effectiveness of the carbon tax by lowering the out-of-pocket cost of solar PV while raising the price of fossil fuels. Under this scheme, the more carbon dioxide you emit, the more you help contribute to its reduction. A carbon tax dedicated to solar creates a strong incentive to shift from fossil fuels to solar PV. Any drag on the economy from the carbon tax would be more than made up by the new economic activity from installing solar PV everywhere.

Germany has succeeded in spurring rapid growth of solar PV by approaching the problem from the opposite direction. Rather than punish carbon emissions with a carbon tax, they reward your new solar installation using a simple mechanism called a *feed-in tariff*—a premium payment for any solar electricity you feed into the electricity grid from your solar PV system. The idea is to pay a sufficiently high price for the solar electricity to turn solar PV into a moneymaker, so that individuals will choose to install their own PV systems for financial gain.

The premium price could be paid for by carbon-tax revenue, but usually is rolled into the electricity bills paid by all utility customers. A study of the effects of the feed-in tariff on German electricity prices found that the net increase was about 1.28 Euro-cents per kilowatt-hour, which represented about 5 percent of the current electricity price.[20]

German utilities paid to PV owners a feed-in tariff as high as 43 Euro-cents (US $.59) per kilowatt-hour of solar PV electricity in 2009, reduced to 29 Euro-cents in 2011, and 17 Euro-cents in 2014. The German feed-in tariff has declined over the years for two reasons: they wanted to encourage action earlier rather than later, and they expected solar PV equipment to get cheaper, and thereby need less subsidy.

Feed-in tariffs have become the most effective incentive program for these reasons:

- Payments are for actual delivered energy, not for equipment that may or may not continue to produce energy once a tax credit for equipment purchase is claimed. This incentive encourages PV owners to maximize their energy output, not just their investment.

- Feed-in tariffs are simple and cheap to administer using utility meter readings.

- The government or utility is not involved in choosing which solar panels must be used or what price should be paid for the equipment; they only pay for generated electricity.
- Steady monthly cash payments are more satisfying to many PV owners than claiming a one-time deduction on their tax form.

The program has turned Germany into a leading installer and manufacturer of solar PV in the world. In 2009 Germany had more than half of the worldwide solar PV market, more than all other countries combined. The United States, where photovoltaics were invented, had only 7 percent.[21] The portion of solar-based electricity (including PV, wind, hydro, and biofuels) in Germany grew from 9 percent to 27 percent in ten years.[22] If these growth rates are sustained, Germany will likely reach its published goal of 80 percent solar-based electricity by 2050.

Germany's 38 gigawatts of installed solar PV (as of 2014) is pushing the limit on how much intermittent solar electricity can be absorbed into the national grid. The German government is supporting the construction of large energy storage systems to allow continued growth in solar PV.[23]

Worldwide, feed-in tariffs have been enacted in sixty-two countries as of 2014,[24] and are reported to have been involved in 75 percent of the PV installations worldwide in 2009.[25]

Feed-in tariffs encourage innovative projects. In 2011, three Indiana companies joined forces in a venture named ET Energy Solutions LLC to take advantage of Indiana's newly enacted feed-in tariffs. They leased 60 acres of open land at the Indianapolis Airport to install and operate 41,000 PV panels, selling the solar electricity to the local utility company. With the 20 cents per kilowatt-hour feed-in tariff, the 11.5 megawatt system provides a profit for the investors. In addition, the Indianapolis Airport Authority gets a steady revenue stream for no investment on their part. Perhaps projects like these are why feed-in tariffs are even recommended for evaluation by the conservative Hoover Institution think tank.[26]

With feed-in tariffs driving PV growth, and new manufacturing techniques developing in response to that growth, the price of solar PV will continue to go down. Once solar PV becomes the cheapest energy source, governments can step aside. The low prices will attract the millions of individuals and organizations that will be needed to carry out such an enormous transition.

Germany has achieved these phenomenal growth rates by creating a program that engages the public, not just energy corporations. Germany now has PV systems of all sizes, from small rooftop systems to utility-scale fields of PV. Each system represents a decision by some individual, from a homeowner to a manager at a utility company, to invest in solar and reap the benefits of the feed-in tariffs. Solar PV is unique in that almost anyone can do it, and that will be a key to its success.

Germany has shown that the right incentives can accelerate change in energy systems. But in the end, they are still just incentives, not directives, and the final choice will be up to individuals.

CHAPTER 25

The Way Forward: Individual Action

The carbon dioxide buildup in our atmosphere is a massive global problem, and yet our world leaders have been unable to reach agreement on an international treaty to control climate change. As we drift into a future of climate catastrophes and pickled oceans, you might wonder, if world leaders can't solve this global problem, who can?

Surprisingly, it seems everyone *except* world leaders can take action. Many individual nations such as Germany have ignored the international gridlock and enacted their own programs to promote solar. In the United States, individual states such as Indiana have ignored the gridlock in Congress and established feed-in tariffs to encourage permanent power from wind and solar.

Even individual cities and counties are taking action. The US Department of Energy named twenty-five cities such as Ann Arbor, Michigan, and Knoxville, Tennessee, as *Solar American Cities* for taking the lead in promoting solar technologies.[1]

For example, the City of Berkeley, California, pioneered innovative solar financing through their Property Assessed Clean Energy (PACE) program. The city offered twenty-year loans for solar PV systems that were paid back through property-tax bills. This program recognized that a solar PV system becomes a permanent part of a home, while adding value to the home. If the property is sold, the new owner continues the payments, while also receiving the benefits of the PV system. Many other jurisdictions now offer PACE programs.

All of these programs are designed to encourage individuals to make the leap to solar. In the final analysis, each decision must be made by a local individual because the circumstances for each solar installation differ.

That's why the top-down approach has had little success. Solar operates at the *micro*-economic level, not the *macro*-economic level. Since everyone uses energy, you are one of several billion contributors to the carbon dioxide problem. Legislation at the international level has little hope of controlling the billions of perpetrators who use energy as part of their daily lives.

If we reexamine the problem from the bottom up, the solution becomes clear. Individuals can apply local solar technologies to their own energy needs, eliminating the carbon dioxide emissions they are directly responsible for. As an energy-using animal, you have a stake in the problem. As a director of energy flows, you have choices for fixing your part of the problem.

Here we can define "individual" as any person responsible for how energy is used at any level. That includes business managers, hospital administrators, government department heads, and army generals, as well as ordinary people.

For most individuals, the best option would be installing solar PV. Most individuals cannot build biofuel plants or offshore wind farms, but individuals can bolt solar panels to their roof. Solar PV gives every individual person and institution the means to fix their own contribution to climate change.

And each decision to install solar PV leads to genuine progress on a global problem. A given installation may seem like a small increment, but it is a real contribution. Unlike carbon trading schemes whose effects are difficult to quantify, the output of PV panels can be measured, so you can know precisely how much carbon you are displacing. And because PV panels last so long, every installation permanently shuts down a small bit of the fossil fuel pipeline. Each newly installed kilowatt brings us closer to the 100 terawatts of PV we will need to reach 100 percent solar in fifty years.

The accumulation of millions of installations will demonstrate to our world leaders an actual working path to a carbon-free future. Hopefully those leaders will become followers, and join the program by removing barriers and enacting incentives to accelerate the transition.

Perhaps you think it naive to believe that individuals making individual choices could possibly bring about any significant change in a major industry such as energy. Nevertheless, we have a precedent.

The rise of organic foods shows that such change is possible. Organic food was not developed or promoted by the giant food and agriculture industries. Rather, organic food became available because consumers demanded it. As consumers learned more about where their foods came from, many would not accept the pesticide residues, soil depletion, and other problems associated with modern food production systems.

Today 53 percent of Americans say they buy some organic food.[2] They are willing to pay a bit more for food that is free of those problems. Initially available only in health food stores, organic food now can be found in every major supermarket, showing how mainstream it has become.

Not only are individuals capable of bringing about the shift from fossil fuels to solar energy, individuals are actually better matched to the task than energy corporations or governments. Here are five reasons why.

1. Individuals Are Distributed, Like the Power Source

Individuals live and work where the sun shines, so they are surrounded by this primary energy source. In contrast, fossil fuels are concentrated in relatively few locations on the planet. Those sites are generally converted into dedicated mining or drilling operations. The raw energy resource must be extracted, refined, and fed into a distribution network to reach consumers.

The raw solar resource is a diffuse energy source spread out over the face of the planet. The solar energy distribution system provided by our Sun happens to reach most of the people who live on the planet.

The space that people occupy for their home and work is space that could be generating electricity with solar PV, if that space has access to the sun. It doesn't take specialized mining and refining equipment to make that energy source useful, just motionless flat panels exposed to the sun.

Also, by generating the power where it is used, the whole system is optimized for maximum efficiency. Electricity generated in a rooftop PV panel can flow directly to its end use on site or be used locally. There are no losses of energy due to transmitting electricity over long distances from a centralized power plant. Such losses are illustrated in Figure 38, "Chain of energy conversions" (page 178).

2. Individuals Know Their Own Energy Patterns

Individuals are in the best position to know their own energy patterns. Individuals can best determine which lights and appliances they use the most, and which are the most inefficient. Individuals know their own driving habits and how they could use a solar-charged electric car.

"The goal of 100 percent solar may be elusive at the international level, but not at the individual level."

Every solar installation is unique, requiring individual onsite investigation of the available roof space or other locations for a PV installation. Individuals can take into account their local utility rates and any tax incentives available in their area.

Individuals are also best situated to combine solar with energy efficiency. Shifting to solar means working the problem from both ends. On the supply side, you try to maximize your production of clean solar electricity, while on the demand side you try to eliminate waste so your precious solar energy can satisfy more of your energy needs.

By combining efficiency measures with a solar PV system that can grow over time, you can push your fossil fuel use down to zero. The goal of 100 percent solar may be elusive at the international level, but not at the individual level.

3. Individuals Can Derive the Most Value

Individuals can reap the most value from their solar electricity. The solar electricity you use on site has an economic value equivalent to the *retail* price of electricity, because it directly displaces electricity you would otherwise have to buy at retail rates. When a company operates a centralized solar power plant that feeds into the power grid, they can only receive about half that as the wholesale price, otherwise the utility company that distributes the power could not cover their costs and make a profit.

Individuals can also assign value to the currently neglected extra benefits that solar electricity carries. Organic food taught us that not all food calories are the same. Similarly, not all energy joules are the same. The long list of significant advantages of solar electricity described in the section titled "The Many Positives of Solar PV" (page 186) includes no carbon dioxide emissions, no air pollution, no water pollution, no long-lived radioactive waste products, and no wars over oil. Economists, following the rules of their trade, cannot rigorously assign monetary values to these advantages, so they leave them out of their economic comparisons of energy alternatives. That leaves solar electricity hugely undervalued, and keeps "clean coal" and nuclear power in contention.

But just because economists do not value solar's advantages does not mean you cannot. An individual has more freedom to evaluate all the pros and cons of their decision to install a PV system. An individual can assign value to their own desire to reduce their carbon footprint, to make the air cleaner, to make the world safer and more secure for their children. An individual can cut through the Gordian Knot that binds economists, and make their decision based on all that they know and all that they feel, not just a narrow monetary accounting.

> People are complex—not mere cost-benefit calculating machines.
>
> —Amory Lovins[3]

4. Individuals Can Care About the Long Term

Individuals are also better at planning for the long term. It's a characteristic of solar PV that it pays out in the long run. There is the large initial investment for the equipment, then several years of electricity production to "pay back" that initial investment. After the payback period, you get years of free electricity, for as long as the panels last. And not just free electricity, but environmentally clean and safe electricity.

Installing solar PV is similar to planting a tree—a lot of up-front effort, some initial satisfaction from the small tree, and a long payout as the tree grows to full size and persists as a landscape feature. The value of solar PV likewise accumulates over time. Every day that it generates electricity adds to its value.

Governments should plan for the long run, but often do not. Governments are run by politicians who are not rewarded for long-term planning, since government budgets are typically for a year, and election cycles are only a few years. In an unusual development, the enormity of the climate-change problem has prompted governments to set carbon-reduction goals ten or fifteen years into the future. But if you examine the details, you'll find they are mostly just goals, and lack specific milestones and budgeting to achieve those goals. Without such specifics, they become little more than wishes.

Corporations also rarely focus on the long term. Market conditions can change rapidly, so long-term commitments must be minimized. The rules of corporate behavior do not reward long-term planning since market analysts require profits every quarter to keep the stock price up.

So neither governments nor corporations are in a position to value permanent power. The fact that solar represents the last energy system humans will ever need to develop does not help a politician get elected or a corporation to show profits. Only individuals can appreciate a permanent power system.

5. Individuals Can Act Now

As an individual, you are in a position to act *now*. Climate scientists continue to pressure governments to act quickly on carbon emissions:

> Waiting for unacceptable impacts to occur before taking action is imprudent because the effects of greenhouse gas emissions do not fully manifest themselves for decades and, once manifested, many of these changes will persist for hundreds or even thousands of years.
>
> —US National Research Council[4]

Of the available alternatives to fossil fuels, only solar-based energies like wind and PV can be deployed widely today. Nuclear proponents say fission reactors are available now, but a full nuclear program will require breeder reactors. Breeder reactors and coal plants with carbon capture are still unproven technologies, and no one can say with certainty when they will be ready or how much they will cost.

But as an individual, you can today call any number of solar contractors, get a firm price, and have a PV system up and running within weeks. Your new system will immediately and permanently displace the annual quota of fossil fuels burned for you at your local power plant. If you plug your system into an electric car or plug-in hybrid, you can immediately and permanently replace your direct use of fossil fuels for transportation. The only question is whether or not you'll choose to do so.

PART V: THE PURSUIT OF PERMANENT POWER

CHAPTER 26

Time to Act

The path forward from our current energy dilemma is clearly visible, based on clean and permanent solar power sources. We are fortunate that the Fossil Fuel Epoch supported the development of permanent power sources like solar PV and wind turbines.

These new technologies present the human race with a unique choice, one we haven't had in the past. We can choose to continue to draw down our stores of fossil-fuel energy and endure their negative consequences, or we can shift to permanent power sources and secure a clean and sustainable future.

We are also fortunate that this path doesn't require an international treaty, because it seems world leaders have abdicated responsibility. Nor do we need the cooperation of the fossil-fuel energy corporations, whose very survival depends on our continuing to use their products.

The choice to shift from fossil fuels to solar-based energy is available at the level of individuals and institutions, by installing solar PV. While politicians dither, you can take direct action to address your part in our global energy problem.

Of course, your PV system alone will not fix global climate change, but you will not be alone. The world is already on an exponential growth path for solar PV. You won't be joining an organization, but joining a movement. As more individuals decide to go solar, we will see a gradual global shift from fossil fuels to solar power.

We don't have to wait for everyone to do it to be successful. Your own success won't depend on others doing the same. Every individual who installs and uses PV power will have successfully shifted their own energy flows from fossil fuels to solar.

In doing so, you join the Solar Epoch. You can personally participate in this historic transformation of the human relationship with energy. The day you switch on your permanent power source marks your entry into a new era.

I Can't Because ...

Perhaps you think you live where it's too cloudy to rely on solar energy. Not everyone lives in sunny California, of course. But the differences in the amount of solar energy available in different parts of the United States are smaller than you might think.

Of the major cities in the U.S., Phoenix, Arizona, receives the most sunshine and Seattle, Washington, the least. If you run the same PV system design through the online PVWatts calculator for these two cities, the results differ by only about 60%.[1] So the worst solar location still has more than half of what the best location has, on an annual basis.

In practical terms, this means Seattle residents will need to install a 60% bigger array to generate electricity equivalent to sunny Arizona. For example, a 4 kilowatt system in Arizona would have to grow to 6.4 kilowatts in Seattle for the same output. Seattle residents will pay more for their PV electricity, but not twice as much, only 60% more.

Perhaps you think you cannot join the Solar Epoch because your rooftop is in the shade, or is unsuitable for a solar installation. Or perhaps you are a renter and don't own a rooftop at all. You can still choose to shift to solar, off site, by joining a community solar project, sometimes called a *solar garden*.

Just as a community garden project lets renters manage their own plots in a shared garden site that gets lots of sun, so do solar gardens permit renters to own a piece of a shared solar PV system installed in a location that gets lots of sun.[2]

For example, the Clean Energy Collective in Carbondale, Colorado, builds centralized PV systems on sunny land they can lease inexpensively. Each member buys a portion of a PV system, and is credited with the solar electricity from their portion as measured on the PV site, including any solar tax credits that may apply. The billing is integrated with the local utility company, so the solar credit directly reduces each member's utility bill. Customers choose how many panels they want to buy, and hold title to the modules purchased. Members get all the benefits of PV ownership without it residing on their own roof.[3]

Solar gardens require the participation of the local electric utility company to manage the electricity and the billing. Solar garden laws to facilitate such participation have been established in Colorado, Delaware, Maine, Massachusetts, Vermont, and Washington, and are pending in several other states.

In England, they call them *solar farms*, and the largest is the Westmill Solar Co-operative outside the town of Swindon. With 5 megawatts of solar PV spread over thirty acres, it generates power for its 1,600 co-op owners.[4]

Here are some other community solar groups that help finance solar projects:

- **Mosaic** is a peer-to-peer solar investment program that helps individuals and organizations that want to go solar obtain low-interest loans, which are funded by other individuals that want to invest in solar.[5]
- **Everybody Solar** helps local non-profit organizations install solar PV to reduce their energy costs and help sustain their community work.[6]
- **RE-volv** accepts donations into a Solar Seed Fund, which is then used to finance solar PV for community projects.[7]

Many electric utilities make it even easier by offering a "green energy" option. For example, Indianapolis Power and Light offers to sell solar-based electricity (primarily from Indiana wind farms and PV) to customers who sign up for their Green Power Option, in which you can choose to get 100 percent, 50 percent, or 25 percent of your electricity from permanent power sources. The premium as of 2014 for Green Power is only $0.002 (two-tenths of a cent) per kwh. A typical residential customer using 1,000 kwh per month at 100 percent Green Power level would pay only an additional $2 on their bill.

Perhaps you think you cannot join the Solar Epoch because you cannot afford the up-front costs, even with the available incentives. Many solar companies will lease to you a PV system that they install and maintain on your roof, with no down payment. Other companies offer a Power Purchase Agreement (PPA), where they install and maintain a PV system on your site and charge you only for the electricity that you use, at a fixed rate established in the contract.

Or you can just wait a few more years. The price of solar PV systems is still going down. If you start saving into a solar fund now, at some point in the future your accumulated savings will meet the descending system price and you will be able to afford it. Since the transition to the Solar Epoch will take fifty years or so, there's no need for everyone to do it at once, even if we could.

Perhaps you think your advanced age prevents you from gaining the full benefit of an investment that lasts thirty years. If so, then buy it as a gift for your children or grandchildren. Solar PV makes a great gift to the younger generations, because they'll be around long enough to reap its full benefits. They're also less likely to be able to afford it, and are more likely to be grateful that someone, finally, is *doing* something about the global warming problems they face. And no one questions the cost effectiveness of a gift.

Buy Your Energy Freedom

Buying a solar PV system also buys you energy freedom. Those without a PV system are captive energy consumers. The development of modern energy systems has turned energy into a commodity that is sold by producers to consumers. The producers are the energy corporations, and the consumers are everyone who uses their energy.

The energy corporations control the fossil fuel energy stores, the extraction process, the refining or conversion process, and the distribution of energy. If you want to participate in this system, you have no choice but to become an energy consumer.

Look at your own life as a captive energy consumer. To drive your car, you must regularly buy fuel. To run your house or business, you must pay your monthly utility bill. All are energy necessities. Since you don't produce that energy yourself, you must buy it, and thereby lock yourself in as an energy consumer.

As a captive consumer, you're subjected to whatever price hikes the energy companies impose. To keep your modern life moving, you have no choice but to pay the higher price.

Being a captive energy consumer also makes you an involuntary global warmer. I doubt you ever chose to contribute to global climate change and ocean acidification, but by playing the role of modern energy consumer, you're forced to be a contributor because those modern energy systems are based largely on fossil fuels.

If you happen to be a captive energy consumer who has also become concerned about climate change, you're in a bind. You can only passively wait for the energy corporations to do something about cleaning up the energy they supply to you. Since they make so much money on fossil fuels, they won't change until the government says they must. But governments have shown little backbone to do so, out of fear of upsetting our fossil fuel-based economy.

You can free yourself from your role as captive energy consumer by installing a solar PV system. When you generate your own electricity from the sun, you no longer contribute to climate change. You can align your energy use with your personal goal of reducing your carbon footprint.

A solar PV system also insulates you from future increases in energy prices. The price you pay for your PV system today determines the average energy cost that it delivers over its lifetime. That "rate" cannot go up because the money has already been spent.

Solar PV amounts to a declaration of independence from the dominant fossil fuel energy system. Don't be surprised if energy corporations try to convince you that it's not in your best interests to generate your own power. Their economic interests would keep you as a captive energy consumer.

Solar PV changes you from a passive energy consumer to an active power producer. You move upstream in the energy supply chain by making your own electricity. The solar electricity you produce is yours to use on site, or to sell into the grid for someone else to use.

Becoming an energy producer changes your attitude toward using energy. As a captive consumer of fossil energy, you must strive to minimize your use of energy, because of the cost burden, and because of the negative impacts that fossil fuels inflict on the world.

As a producer of solar electricity, you instead strive to maximize your use of your own energy. The energy you produce is clean and safe, with no negative consequences. The more you use, the more fossil fuel energy you displace. The act of using more energy becomes good for the world, not bad.

With solar PV, you not only become a producer, you become a *primary* producer by converting a raw natural resource into a useful product. If you have a solar PV system, you become a primary producer because you convert raw sunlight into useful electricity. That gives you a fundamental role in the economy, a role that will have value even in a recession, because we will always need energy.

Regardless of the economics, I derive a great deal of personal satisfaction from using my own energy. There's nothing like driving my solar-charged electric car past a gas station that just raised its prices again. That feeling of self-reliance is priceless.

Advance the Human Race

Consider where your choice to go solar will place you in the history of the world.

You can choose to do nothing, and continue to buy fossil-fuel-based energy as you always have. Then future generations, coping with the effects of accumulated carbon dioxide, will look back to see how they got into that situation, and blame those who did nothing to correct the problems when they had a chance to do so.

You can instead choose to examine the energy flows that you direct and figure out how to convert them to permanent power sources. Every energy-using animal has that choice.

If our collective volition manages to carry out a global shift to solar power sources, then you'll be honored as a pioneer. Since the Solar Epoch will last indefinitely on permanent power sources, you'll be among the first of hundreds or thousands of generations who use solar energy in the future. Ours will be the generation to start the human race down this path.

Not many generations have the opportunity to change the fundamental human relationship with energy. The last time it happened was in the early 1800s, around the time of Thomas Jefferson. The antique tall clock in my living room dates from that time. My clock, built in 1820, witnessed the rise of the Fossil Fuel Epoch and the development of all the modern energy systems we have come to depend on. It was present when the first electric lights were turned on, the first refrigerator moved in, and the first smart phone showed up.

When the clock was moved to my home, it also witnessed the start of the Solar Epoch. My home is largely solar powered, and thus participates in the shift from fossil energy stores to solar power sources. The big clock fits right in, since it too is solar powered, by way of photosynthesis making food to power the muscles in my arm to wind it. As long as someone is around to wind it, this old clock, which witnessed the entire Fossil Fuel Epoch, will carry on indefinitely into the Solar Epoch.

My example shows you how easy it is to join the Solar Epoch. Just start living on permanent power sources instead of energy stores. You'll do yourself and human society a great good. As more individuals decide to switch, and as governments and other institutions decide to support it, we will carry out the gradual shift to permanent solar power so that one day we can relegate fossil fuel equipment to museums.

My clock survived the Fossil Fuel Epoch and came out the other side into the Solar Epoch. It is not yet clear if human civilization will succeed in doing the same.

We have before us an opportunity to advance civilization to the next level. The coming decades will reveal if we act on that opportunity, and step into the sunshine.

About Robert Arthur Stayton

Robert Arthur Stayton lives and writes in Santa Cruz County, California. He has a masters degree in physics and has taught college courses in physics, energy, and solar energy.

Robert received his bachelor's degree in physics at the University of Colorado in Boulder in 1974. He was then awarded a scholarship for graduate study in physics at the University of California, Santa Cruz (UCSC). His research interests there included superconductivity and solar energy. His interest in solar energy led him to develop and teach a course entitled *Energy* for the UCSC Environmental Studies department. Robert decided to pursue a career in teaching energy rather than researching physics, so he completed the requirements for a master's degree in physics and began teaching at Cabrillo College in Aptos, California.

At Cabrillo College, Robert taught courses on The Fundamentals of Solar Energy, Solar Home Design, and Wind Energy, as well as physics courses. Claire Biancalana, Dean of Occupational Education at Cabrillo, said that the solar department "has an extraordinary instructor in Bob Stayton". Robert also taught courses on Alternative Energy for the Environmental Studies Department at UC Santa Cruz.

Robert also actively participated in community energy activities. He co-founded Energy Action, a research group that received a grant from the California Energy Commission to develop a baseline energy profile for Santa Cruz County. He also co-founded Energy Future Santa Cruz, a nonprofit organization that received a grant from the National Science Foundation to develop an energy plan for Santa Cruz County using input from government agencies, community organizations, schools, and citizens. He also served as Energy Advisor for Santa Cruz County

Supervisor Pat Liberty and served on the Energy Advisory Committee for the City of Santa Cruz.

Robert's interest in writing developed while still a graduate student in physics at UCSC. After taking a course in Science Writing, he was appointed a Science Writer Intern for the California State Assembly Office of Research in Sacramento. There he developed a white paper on the *Prospects for the Use of Solar Energy in California* that was circulated to state legislators. Several months later the legislature passed AB1558 that established the most advanced solar tax credit in the U.S. One state department head wrote to David Saxon, President of the University of California at the time, to praise the white paper, saying "For a scientist to be able to communicate in language a lay person can understand is a rare quality. Mr. Stayton demonstrates this gift in this very fine publication."

Robert also completed the Graduate Program in Science Communication at UC Santa Cruz, and has had articles published in *Popular Science* and *Science Notes*.

In 1997, Robert and his wife built a passive solar home in Santa Cruz County and outfitted it with an off-grid solar photovoltaic system. He has been living with solar energy since then, always looking for new ways to apply solar in his daily life. He drives a solar-charged Plug-in Prius, heats his water with a solar water heating system, and bakes his bread in his solar oven. He has served as host to hundreds of people who have toured his home to see his solar efforts.

Acknowledgments

The person most deserving of acknowledgment and thanks is my wife, Mary Tsalis. She has patiently supported me for over fifteen years as I developed and wrote this book, a project that seemed to have no end in both our minds. She provided crucial feedback from her repeated readings of the manuscript, and helped shape the content to fit my intended general audience. I could not have done this book without her.

I'd like to give special thanks to John Wilkes, who as head of the Graduate Program in Science Communication at the University of California Santa Cruz nurtured my writing career and convinced a science nerd that he could be a writer.

I want to thank my reviewers for contributing their time to read and thoughtfully comment on my book: Mike Arenson, Hal Aronson, Jonathan Berkey, Len Beyea, Douglas Brown, Candace Calsoyas, Peter Cooney, Francis de Winter, Jennie Dusheck, Mark Forry, Matthew Gilbert, Sandora Hedrich, Daniel Hirsch, Joe Jordan, Andrea Lackides, Athena Lackides, Nicholas Lackides, Doug McKenzie, Margaret Payne, Ron Pomerantz, Sarah Rabkin, Donna Riordan, Michael Riordan, Barbara Stayton, Mike Stayton, Tom Stayton, Ron Swenson, Joe Symons, Jean Thomas, John Wilkes, Matt Wood, and Eric Youngren. They all helped improve this book.

I'd like to thank Berkeley Kauffman for his research work, Sam Case for his fact checking and copy editing, Sandy Bell of Sandy Bell Design for her cover and interior design work, and Joe Shaw at Cypress Press for his editing work. My illustrator Todd Sallo deserves special thanks for his extra efforts to research and design each of the illustrations for maximum effect.

It was my privilege to work with Marilyn McGuire to market my book. Her experience and connections helped me avoid countless pitfalls. I especially want to thank her for the extra time she committed to the project because she believe in it. I also received valuable marketing advice from Cynthia Frank of Cypress Press and Michael Riordan of Michael Riordan Productions.

Finally, I have to acknowledge John Muir—not the naturalist, but the mechanic (1918–1977). John Muir wrote *How to Keep Your Volkswagen Alive; A Manual of Step-By-Step Procedures for the Compleat Idiot.* In my younger days, I used that book to successfully rebuild six Volkswagen engines. The fact that I had no previous experience in car mechanics shows how effective the book was. I have long harbored in the back of my mind the desire to write a book that made a complex subject easy for anyone to understand. That's what I've tried to do with the subject of energy in *Power Shift*.

Notes and References

Chapter 1: Energy Defines Us

1. The size of the equivalent bubble of carbon dioxide is computed as follows. The 2010 worldwide emissions of carbon dioxide were 33,615,389,000 metric tons (Wikipedia article "List of countries by carbon dioxide emissions," accessed 13 August 2014). A ton of carbon dioxide at standard temperature and pressure has a volume of 556.2 cubic meters. Each day's emission was therefore 51.2 billion cubic meters in volume, which would fit in a bubble 2.304 km in radius, 4.6 km in diameter (1.609 km per mile), or 2.861 miles in diameter. The value is rounded up in the text.

2. *Ocean Acidification: State of the Science Fact Sheet*, US National Oceanic and Atmospheric Administration, January 2013, http://www.noaa.gov/factsheets/new%20version/SoS%20Fact%20Sheet_Ocean%20Acidification%2020130306%20Final.pdf, (accessed 14 August 2014).

3. Elizabeth Kolbert, "The Acid Sea," *National Geographic*, April 2011.

4. The concentration of carbon dioxide changed from 280 parts per million prior to the Industrial Revolution to 400 parts per million in 2014. Wikipedia, "Carbon dioxide in Earth's atmosphere." The number for the volume of atmosphere extends to 100 kilometers in altitude, which includes the 99.9999 percent of the atmosphere.

5. *Ocean Acidification: State of the Science Fact Sheet*, US National Oceanic and Atmospheric Administration, January 2013, http://www.noaa.gov/factsheets/new%20version/SoS%20Fact%20Sheet_Ocean%20Acidification%2020130306%20Final.pdf, (accessed 14 August 2014). The pH scale is a measure of the concentration of chemically reactive hydronium ions dissolved in water. Because the pH scale is logarithmic, reducing the pH by 1 increases the concentration of ions by ten times. The extra carbon dioxide has reduced the oceans' pH by 0.11, which corresponds to a 30% increase in ion con-

centration. For more on pH, see http://pmel.noaa.gov/co2/story/A+primer+on+pH.

6. Emission rates between 1997 and 2012 rose from 24.4 to 34.5 billion metric tonnes. Source: Jos G.J. Olivier, et al., *Trends in global CO2 emissions: 2013 Report* (PBL Netherlands Environmental Assessment Agency, The Hague, 2013), PBL publication number: 1148 JRC Technical Note number: JRC83593 EUR number: EUR 26098 EN.

Chapter 2: How We Became the Energy-using Animal

1. If you are wondering about the senses of smell, taste, and touch, they all require contact. In the case of smell, an actual molecule has to travel through the air and lodge in your nose to be detected. Likewise with the sense of taste on the tongue. The sense of touch obviously requires contact. None of these detect action at a distance through the mechanism of energy transfer.

2. We have no machines that can directly convert chemical energy to kinetic energy. Heat engines that run on chemical fuels first burn the fuel to make heat, and then some of that heat is converted to motion. Fuel cells convert chemical energy to electricity that can then be converted to motion with an electric motor.

3. Food energy intake ranges from 2,000 to 3,000 kilocalories (1 kilocalorie = 1 food Calorie) per day, depending on age, gender, and activity level. A gallon of gasoline contains 132 million joules, or 31,549 kcal. A cup is one sixteenth of a gallon, so a cup of gasoline contains 1,972 kcal. So the gasoline equivalent to food intake would range from 1.01 to 1.52 cups (half-pints) of gasoline per day.

4. The energy from fossil fuels and electricity that is used in farming, processing, and shipping food does not end up in the food itself.

5. Chemotrophs are life forms that do not need solar energy, but can derive their energy from chemicals in their environment. These include certain bacteria that feed on hydrogen sulfide emitted by some deep-sea thermal vents. Giant tube worms living by the vents can use such bacteria in a symbiotic relationship to make their food. These are not advanced life forms.

6. Movies are motion without matter, and so movies have no kinetic energy.

7. Wikipedia contributors, "Arrowhead," Wikipedia, The Free Encyclopedia, http://en.wikipedia.org/wiki/Arrowhead (accessed 22 July 2014).

Chapter 3: The Wood Energy Epoch

1. The term *archaic humans* has no single agreed-upon definition, but generally includes Neanderthals and other anatomically different human species. *Nature* 505, 32–34 (02 January 2014).

2. Jean Gimpel, *The Medieval Machine: The Industrial Revolution of the Middle Ages* (Penguin, New York, 1975), pg. 79.

3. Vaclav Smil, *Energy in Nature and Society: General Energetics of Complex Systems* (The MIT Press, 2008) pg. 372.

4. Vaclav Smil, *Energy in Nature and Society: General Energetics of Complex Systems* (The MIT Press, 2008) pg. 159.

5. Rotational kinetic energy is not something you find in nature on the scale of human existence. Our rotating planet has natural rotational kinetic energy, but we cannot tap into it.

6. Karl Marx, *Capital: Critique of Political Economy* (France, 1867).

7. Physicists use the more formal term *gravitational potential energy* to describe the energy associated with an elevated weight. Here *potential* means the energy is not readily apparent but is capable of performing some task when converted to another form such as kinetic energy. The chemical energy stored in a battery or our food could also be called chemical potential energy.

8. The Domesday Book Online, http://www.domesdaybook.co.uk/

9. Julian Hatcher, *Hatcher's Notebook* (The Stackpole Company, Harrisburg, PA, 1962).

10. Strictly speaking, some crossbows could penetrate some armor, and some catapults could penetrate some castle walls, but guns and cannons were irresistible, making both armor and castle walls obsolete.

Chapter 4: The Fossil Fuel Epoch

1. "Coal mining; Ancient use of outcropping coal," *Encyclopedia Britannica*, http://www.britannica.com/EBchecked/topic/122975/coal-mining/81626/History.

2. T. S. Ashton and Joseph Sykes, *The Coal Industry of the Eighteenth Century*, 2nd edition (Manchester, England, Manchester University Press, 1964).

3. Jenny Uglow, *The Lunar Men: Five Friends Whose Curiosity Changed the World* (London: Faber & Faber), pg. 376.

4. Within the cases bounded by the Uncertainty Principle in quantum mechanics.

5. The heat energy calorie differs from the food Calorie, which is actually equal to 1,000 heat calories and is usually spelled with a capital "C." Although derived from metric units, the calorie itself is not included in the official metric system, perhaps because of this confusion.

6. *Annual Energy Review 2009*, Energy Information Administration, US Department of Energy, DOE/EIA-0384(2009), August 2010.

7. Electrical energy is usually expressed in units of *kilowatt-hours,* abbreviated kwh. A kilowatt-hour of energy is equivalent to one kilowatt of power applied for one hour. Note that this is not a kilowatt per hour, but a kilowatt *times* an hour. Since a kilowatt is 1,000 joules per second, and an hour contains 3,600 seconds, multiplying the two together shows that one kwh is equal to 3.6 million joules.

8. Wikipedia contributors, "Nitroglycerin," Wikipedia, The Free Encyclopedia, http://en.wikipedia.org/wiki/Nitroglycerin (accessed 5 August 2014).

9. Heather L. Tierney, et al., "Experimental demonstration of a single-molecule electric motor," *Nature Nanotechnology,* vol 6, 625–629 (2011) doi:10.1038/nnano.2011.142.

10. Lightning is a natural source of electricity, but it cannot be controlled by humans.

11. *International Energy Outlook 2011*, Energy Information Administration, US Department of Energy, September 2011.

12. Each cycle of an internal combustion engine takes two rotations of the engine. When an engine runs at a relatively low 1200 revolutions per minute (rpm), that's 600 cycles per minute, or 10 per second. At 3600 rpm, each cylinder is firing 30 times per second.

13. "Fortune Global 500, 2009", http://money.cnn.com/magazines/fortune/global500/2005/index.html

14. *History of the American West, 1860-1920: Photographs from the Collection of the Denver Public Library*, United States Library of Congress, http://www.loc.gov/teachers/classroommaterials/connections/hist-am-west/history.html.

15. *Statistical Abstract of the United States: 2012*, United States Census Bureau, Table 620: Employment by Industry: 2000 to 2010.

16. Vaclav Smil, *Energy in Nature and Society: General Energetics of Complex Systems* (The MIT Press, 2008) pg. 377.

Chapter 5: Fossil Fuels and Heat Engines Advanced the Human Race

1. Household ice boxes that were periodically refilled with blocks of ice did not become prevalent until the mid-1800s.

2. Vaclav Smil, *Energy in Nature and Society: General Energetics of Complex Systems* (The MIT Press, 2008) pg. 380

Chapter 6: Modern Energy Consumers

1. Richard Feynman, Robert B. Leighton, Matthew Sands, *The Feynman Lectures on Physics*, 2011 edition (Basic Books). Originally published in 1963.

2. The number of energy servants is computed as follows: US Energy Information Administration Annual Energy Review 2010 says total US energy use was 98,003 trillion BTU per year, where 1 BTU = 1055 joules. US population in 2010 was 308.7 million people. That's 98,003 trillion BTU times 1055 joules per BTU divided by 308.7 million people = 334.9 GJ per person per year, or 10,620 joules per second (watts). Energy requirement for the average man is 2700 kcal/day, and 2100 for women. Assume half men and half women servants, or 2400 kcal / day average. Since 1 kcal = 4184 joules, food energy flow is 2400 kcal/day times 4184 joules/kcal divided by 24hr/day divided by 3600 second/hr = 116 joules/second (watts). So 10,620 watts divided by 116 watts per servant = 91 servants that could be supported on the equivalent food energy. Fossil fuels in that mix amounted to 81,420 trillion BTU, or 83%. So if we only use servants powered by fossil fuels, that would be about 75 servants per person.

Chapter 7: Fossil Fuels Put Us in a Bind

1. John S. Steinhart and Carol E. Steinhart, "Energy Use in the U.S. Food System," *Science*, 19 April 1974: 307-316.

2. David Pimentel, et al., "Reducing Energy Inputs in the US Food System," *Human Ecology* (2008) 36:459–471. The report states that the average US diet of 2,547 kcal/day is supported by 2,000 liters of oil equivalent per year. Crude oil energy content is 37 MJ/liter, so that's 2000 x 37 MJ = 74 GJ of oil equivalent. Food energy need is 2547 kcal/day x 4184 J/kcal x 365 days = 3.89 GJ of food energy per year. Ratio of 74 GJ to 3.89 GJ is 19 to 1.

3. Intergovernmental Panel on Climate Change, *Climate Change 2013: The Physical Science Basis. Contribution of Working Group I to the Fifth Assessment Report of the Intergovernmental Panel on Climate Change* [Stocker, T.F., D. Qin, G.-K. Plattner, M. Tignor, S.K. Allen, J.

Boschung, A. Nauels, Y. Xia, V. Bex and P.M. Midgley (eds.)] (Cambridge University Press, Cambridge, United Kingdom and New York, NY, USA).

4. *State of the Climate: Global Analysis - Annual 2013*, National Climatic Data Center, US National Oceanic and Atmospheric Administration, 2013, http://www.ncdc.noaa.gov/sotc/global/2013/13, (accessed 13 August 2014).

5. Intergovernmental Panel on Climate Change, 2013: *Summary for Policymakers. In: Climate Change 2013: The Physical Science Basis. Contribution of Working Group I to the Fifth Assessment Report of the Intergovernmental Panel on Climate Change* [Stocker, T.F., D. Qin, G.-K. Plattner, M. Tignor, S.K. Allen, J. Boschung, A. Nauels, Y. Xia, V. Bex and P.M. Midgley (eds.)] (Cambridge University Press, Cambridge, United Kingdom and New York, NY, USA). pg. 4.

6. *Lighting the way: Toward a sustainable energy future*, InterAcademy Council, Amsterdam, October 2007.

7. Intergovernmental Panel on Climate Change, *Climate Change 2013: The Physical Science Basis. Contribution of Working Group I to the Fifth Assessment Report of the Intergovernmental Panel on Climate Change* [Stocker, T.F., D. Qin, G.-K. Plattner, M. Tignor, S.K. Allen, J. Boschung, A. Nauels, Y. Xia, V. Bex and P.M. Midgley (eds.)] (Cambridge University Press, Cambridge, United Kingdom and New York, NY, USA). pg. 14.

8. John Bragg, "Acidification Is 'Fundamentally Altering' Oceans," *Coastal and Estuarine Research Federation Newsletter*, September 2009.

9. N. Bednaršek, et al., "*Limacina helicina* shell dissolution as an indicator of declining habitat suitability owing to ocean acidification in the California Current Ecosystem," *Proceedings of the Royal Society B*, Vol. 281, No. 1785, 22 June 2014.

10. *Workshop Report, IPCC Workshop on Impacts of Ocean Acidification on Marine Biology and Ecosystems*, Intergovernmental Panel on Climate Change, Okinawa, Japan 17–19 January 2011.

11. A. D. Rogers and D.d'A. Laffoley, *International Earth system expert workshop on ocean stresses and impacts. Summary report*, International Programme on the State of the Ocean (IPSO) 2011, Oxford, pg. 5.

12. Scott C. Doney, "The Growing Human Footprint on Coast and Open-Ocean Biogeochemistry," *Science* vol 328, 18 June 2010. pp 1512–1516.

13. *Workshop Report, IPCC Workshop on Impacts of Ocean Acidification on Marine Biology and Ecosystems*, Intergovernmental Panel on Climate Change, Okinawa, Japan 17–19 January 2011.

14. Elizabeth Kolbert, "The Acid Sea," *National Geographic*, April 2011.

15. George A. Olah, Alain Goeppert, and G.K. Surya Prakash,*Beyond Oil and Gas: The Methanol Economy*, Second updated and enlarged edition (Wiley-VCH Verlag GmbH & Co. KGaA, Weinhiem, 2009). pg. 58.

16. Travis Bradford, *Solar Revolution: The Economic Transformation of the Global Energy Industry* (The MIT Press, Cambridge MA, 2006) , pg. 57.

17. Gretchen Goldman, et al., *Toward an Evidence-Based Fracking Debate: Science, Democracy, and Community Right to Know in Unconventional Oil and Gas Development*, Union of Concerned Scientists, October 2013. pg. 3.

18. *Annual energy outlook 2013. With projections to 2040*, Energy Information Administration (EIA), U.S. Department of Energy. DOE/EIA-0383(2013), http://www.eia.gov/forecasts/ aeo/pdf/ 0383%282013%29.pdf (accessed August 21, 2014).

19. R. A. Alvarez, et al., "Greater focus needed on methane leakage from natural gas infrastructure," *Proceedings of the National Academy of Sciences* 109:6435–6440, 2012, http://www.pnas.org/content/109/17/6435.

20. Intergovernmental Panel on Climate Change, *Climate Change 2013: The Physical Science Basis. Contribution of Working Group I to the Fifth Assessment Report of the Intergovernmental Panel on Climate Change* [Stocker, T.F., D. Qin, G.-K. Plattner, M. Tignor, S.K. Allen, J. Boschung, A. Nauels, Y. Xia, V. Bex and P.M. Midgley (eds.)] (Cambridge University Press, Cambridge, United Kingdom and New York, NY, USA).

21. Anna Karion, et al., "Methane emissions estimate from airborne measurements over a western United States natural gas field," *Geophysical Research Letters* Volume 40, Issue 16, pages 4393–4397, 28 August 2013.

22. Gretchen Goldman, et al., *Toward an Evidence-Based Fracking Debate: Science, Democracy, and Community Right to Know in Unconventional Oil and Gas Development*, Union of Concerned Scientists, October 2013.

23. Wikipedia contributors, "Exemptions for hydraulic fracturing under United States federal law," Wikipedia, The Free Encyclopedia, http://en.wikipedia.org/wiki/Exemptions_for_hydraulic_fracturing_under_United_States_federal_law (accessed 5 August 2014).

24. Tim McMahon, *Historical Crude Oil Prices (Table)*, InflationData.com, http://inflationdata.com/Inflation/Inflation_Rate/Historical_Oil_Prices_Table.asp, accessed 23 July 2014.

25. Tim McMahon, *Historical Crude Oil Prices (Table)*, InflationData.com, http://inflationdata.com/Inflation/Inflation_Rate/Historical_Oil_Prices_Table.asp, accessed 23 July 2014.

26. *Annual Energy Review 2010*, Energy Information Administration, US Department of Energy, DOE/EIA-0384(2010), October 2011. pg. 179.

27. *2010 Key World Energy Statistics*, International Energy Agency, Paris, France, http://www.iea.org

28. *Global Trends 2025: A Transformed World*, U.S. National Intelligence Council, November 2008. www.dni.gov/nic/NIC_2025_project.html. pg. 45.

29. *Global Trends 2025: A Transformed World*, U.S. National Intelligence Council, November 2008. www.dni.gov/nic/NIC_2025_project.html. pg. 45.

30. *World Oil Transit Chokepoints*, Energy Information Administration, US Department of Energy, 2011.

31. Jenifer Piesse and Colin Thirtle, "Three bubbles and a panic: An explanatory review of recent food commodity price events," *Food Policy* 34(2): 11 (2009).

32. Oil data from "Europe Brent Spot Price FOB," US Energy Information Administration, Petroleum and Other Liquids, http://www.eia.gov/dnav/pet/hist/LeafHandler.ashx?n=PET&s=RBRTE&f=M, retrieved 22 July 2014. Food data from "FAO Food Price Index," Food and Agriculture Organization of the United Nations, http://www.fao.org/worldfoodsituation/foodpricesindex/en/, retrieved 22 July 2014. Note: these monthly oil prices show greater highs and lows than the annual average prices shown in Figure 22, "Crude oil price history (in constant 2014 dollars)" (page 88), where the highs and lows are smoothed by averaging.

33. Jenifer Piesse and Colin Thirtle, "Three bubbles and a panic: An explanatory review of recent food commodity price events," *Food Policy* 34(2): 11 (2009).

34. H. Charles J. Godfray, et al., "Food Security: The Challenge of Feeding 9 Billion People," *Science* Vol 327, 12 February 2010.

35. *Air Toxics from Motor Vehicles*, US Environmental Protection Agency, EPA Fact Sheet EPA400-F-92-004, August 1994.

36. David C. Holzman, "AIR POLLUTION: Mercury Emissions Not Shrinking as Forecast," *Environmental Health Perspective*, May 2010, 118(5): A198.

37. Alex Gabbard, "Coal Combustion: Nuclear Resource or Danger?," *Oak Ridge National Laboratory Review* Vol. 26, Nos. 3 & 4, 1993.

38. "Ambient (outdoor) air quality and health," United Nations World Health Organization Fact sheet No. 313, March 2014, http://www.who.int/mediacentre/factsheets/fs313/en/

39. Michael Marshall, "Poisonous Ingredients," *New Scientist*, 4 June 2011, pg. 10.

40. *Oil in the Sea III: Inputs, Fates, and Effects*, US National Research Council, National Academies Press, 2003.

41. *Oil in the Sea III: Inputs, Fates, and Effects*, US National Research Council, National Academies Press, 2003.

42. *Oil in the Sea III: Inputs, Fates, and Effects*, US National Research Council, National Academies Press, 2003.

43. Emily S. Bernhardt and Margaret A. Palmer, "The environmental costs of mountaintop mining valley fill operations for aquatic ecosystems of the Central Appalachians," *Annals of the New York Academy of Sciences* 1223 (2011) pp 39-57.

44. Alan Neuhauser, "Study: Chemical in W.Va. Spill More Toxic Than Thought," *U.S. News and World Report* 10 July 2014.

45. *Waxman, Markey, and DeGette Investigation Finds Continued Use of Diesel in Hydraulic Fracturing Fluids*, U.S. House of Representatives Committee on Energy & Commerce, 31 January 2011.

46. U.S. Environmental Protection Agency, *Evaluation of Impacts to Underground Sources of Drinking Water by Hydraulic Fracturing of Coalbed Methane Reservoirs* (June 2004) (EPA 816-R-04-003) at 4-11.

47. "Pavillion, Wyoming Groundwater Investigation: January 2010 Sampling Results and Site Update," US Environmental Protection Agency, August 2010, http://www2.epa.gov/sites/production/files/documents/PavillionWyomingFactSheet.pdf

48. Melissa M. Ahern, et al., "The association between mountaintop mining and birth defects among live births in central Appalachia, 1996–2003," *Environmental Research* Volume 111, Issue 6, August 2011, pages 838–846.

49. *CO2 Emissions from Fuel Combustion: Highlights*, 2010 Edition, International Energy Agency, Paris, France.

50. Solomon, S., D. Quin, M. Manning, Z. Chen, M. Marquis, K.B. Averyt, M. Tignor and H.L. Mikller (eds.), *Climate Change 2007: The Physical Science Basis. Contribution of Working Group I to the Fourth Assessment Report of the Intergovernmental Panel on Climate Change* (Cambridge University Press, United Kingdom and New York, NY, 2007). , pg. 803.

Chapter 8: Crucial Decision Point on Energy

1. In the five-year period from 1990 to 1995, carbon emissions grew by 4 percent, from 20.9 to 21.8 billion tons per year. In the five-year period from 2005 to 2010, carbon emissions grew by 11 percent, from 27.2 to 30.3 billion tons per year. [*CO2 Emissions from Fuel Combustion:Highlights*, 2013 Edition, International Energy Agency, Paris, France.]

Chapter 9: Eliminate Modern Energy

1. Vaclav Smil, *Energy in Nature and Society: General Energetics of Complex Systems* (The MIT Press, 2008) pg. 305

Chapter 10: Clean Coal

1. *IPCC Special Report on Carbon Dioxide Capture and Storage*, prepared by Working Group III of the Intergovernmental Panel on Climate Change [Metz, B., O. Davidson, H. C. de Coninck, M. Loos, and L. A. Meyer (eds.)] (Cambridge University Press, Cambridge, United Kingdom and New York, NY, USA).

2. Mark Z. Jacobson and Mark A. Delucchi, "Providing all global energy with wind, water, and solar power, Part I: Technologies, energy resources, quantities and areas of infrastructure, and materials," *Energy Policy* (2010) doi:10.1016/j.enpol.2010.11.040.

3. *IPCC Special Report on Carbon Dioxide Capture and Storage*, prepared by Working Group III of the Intergovernmental Panel on Climate Change [Metz, B., O. Davidson, H. C. de Coninck, M. Loos, and L. A. Meyer (eds.)] (Cambridge University Press, Cambridge, United Kingdom and New York, NY, USA).

4. *IPCC Special Report on Carbon Dioxide Capture and Storage*, prepared by Working Group III of the Intergovernmental Panel on Climate Change [Metz, B., O. Davidson, H. C. de Coninck, M. Loos, and L. A. Meyer (eds.)] (Cambridge University Press, Cambridge, United Kingdom and New York, NY, USA).

5. Antony Froggatt and Glada Lahn, *Sustainable Energy Security*, Lloyd's 360° Risk Insight, London, 2010.

6. Global oil production in 2013 was 90.5 million barrels per day [http://www.eia.gov/forecasts/steo/pdf/steo_full.pdf]. Global carbon dioxide emissions in 2013 were 36 billion tons [http://cdiac.ornl.gov/GCP/carbonbudget/2013/]. If the carbon dioxide is compressed to its maximum supercritical level, then it has a density of 0.469 gm/cm^3 [http://en.wikipedia.org/wiki/Supercritical_fluid]. At that pressure, the volume of carbon dioxide would be 1,323 million barrels per day. That amounts to 14.7 times the daily volume of oil extracted worldwide.

7. *IPCC Special Report on Carbon Dioxide Capture and Storage*, prepared by Working Group III of the Intergovernmental Panel on Climate Change [Metz, B., O. Davidson, H. C. de Coninck, M. Loos, and L. A. Meyer (eds.)] (Cambridge University Press, Cambridge, United Kingdom and New York, NY, USA).

8. "Carbon storage technology is far from ready, utility execs warn," *E&ENews*, June 17, 2009.

Chapter 11: Nuclear Energy

1. Larry Parker and Mark Holt, *Nuclear Power: Outlook for New U.S. Reactors*, US Congressional Research Service, Order Code RL33442, March 2007.

2. Stewart Brand, *Whole Earth Discipline: Why Dense Cities, Nuclear Power, Transgenic Crops, Restored Wildlands, and Geoengineering Are Necessary* (Penguin Books, Reprint edition, September 28, 2010).

3. The ratio of 3 million is theoretical. In current nuclear reactors, only a small fraction of the uranium-235 atoms can actually be fissioned before the byproducts contaminate the process.

4. *International Energy Outlook 2011*, Energy Information Administration, US Department of Energy, September 2011.

5. A few nuclear power sites also distribute some of their waste heat by piping hot water to nearby buildings. See "District heating," Wikipedia, The Free Encyclopedia, http://en.wikipedia.org/wiki/District_heating (accessed 14 August 2014).

6. Nathan S. Lewis, "Powering the Planet," *Engineering & Science* No. 2, 2007.

7. *Energy [R]evolution: A Sustainable World Energy Outlook*, Greenpeace International and European Renewable Energy Council, 3rd edition, June 2010.

8. Nathan S. Lewis, "Powering the Planet," *Engineering & Science* No. 2, 2007.

9. Wikipedia contributors, "Breeder reactor," Wikipedia, The Free Encyclopedia, http://en.wikipedia.org/wiki/Arrowhead (accessed 5 August 2014).

10. Joshua M. Pearce, "Thermodynamic limitations to nuclear energy deployment as a greenhouse gas mitigation technology," *International Journal of Nuclear Governance, Economy and Ecology* Vol. 2, No. 1, 2008, pg. 113.

11. *Enformable Nuclear News*, 18 October 2011, http://enformable.com/2011/10/over-7000-tons-of-contaminated-rice-straw-unable-to-be-stored-farmers-stuck-with-piles-of-clean-straw-and-piles-of-contaminated-straw/

12. *Reuters*, 28 September 2011, http://www.reuters.com/article/2011/09/28/japan-nuclear-waste-idUSL3E7KS0GM20110928

13. "Japan begins building Fukushima ice wall," *BBC News*, 2 June 2014, http://www.bbc.com/news/world-asia-27669393 (accessed 10 September 2014).

14. The curie is a measure of radioactivity, roughly equivalent to the activity of 1 gram of the radium-226, a substance studied by Marie and Pierre Curie.

15. James C. Warf, *All Things Nuclear*, Second Edition (Figueroa Press, Los Angeles, CA, 2004).

16. Wikipedia contributors, "Goiânia accident," Wikipedia, The Free Encyclopedia, http://en.wikipedia.org/wiki/Goi%C3%A2nia_accident (accessed 9 September 2014).

17. *Survey of National Programs for Managing High-Level Radioactive Waste and Spend Nuclear Fuel: A Report to Congress and the Secretary of Energy*, U.S. Nuclear Waste Technical Review Board. October 2009.

18. David Biello, "Presidential Commission Seeks Volunteers to Store U.S. Nuclear Waste," *Scientific American*, July 2011.

19. Wikipedia contributors, "Thorium-based nuclear power," Wikipedia, The Free Encyclopedia, http://en.wikipedia.org/wiki/Thorium-based_nuclear_power (accessed 16 October 2014).

20. Wikipedia contributors, "Thorium-based nuclear power," Wikipedia, The Free Encyclopedia, http://en.wikipedia.org/wiki/Thorium-based_nuclear_power (accessed 16 October 2014).

21. Andrew T. Nelson, "Thorium: Not a near-term commercial nuclear fuel," *Bulletin of the Atomic Scientists* 68(5) 33-44, 2012.

22. Phil McKenna, "Is the 'Superfuel' Thorium Riskier Than We Thought?," *Popular Mechanics*, 5 December 2012, http://www.popular-mechanics.com/science/energy/nuclear/is-the-superfuel-thorium-riskier-than-we-thought-14821644, accessed October 2014.

23. World Nuclear Association, *Nuclear Power in India*, September 2014, http://www.world-nuclear.org/info/Country-Profiles/Countries-G-N/India/

24. Wikipedia contributors, "India's three-stage nuclear power programme," Wikipedia, The Free Encyclopedia, http://en.wikipedia.org/wiki/India's_three-stage_nuclear_power_programme (accessed 16 October 2014).

Chapter 12: Solar Energy

1. In Sacramento California, the measured amount of solar energy on a south-facing surface tilted up 23 degrees received 5.5 kilowatt-hours per square meter per day averaged over a year [US National Renewable Energy Laboratory, "Solar Radiation Data Manual for Flat-Plate and Concentrating Collectors"]. That converts to 671 million joules per square foot per year. One gallon of gasoline contains 132 million joules. So one square foot of roof provides the equivalent of 5.1 gallons of gasoline per year.

2. Stacy C. Davis, et al., *Transportation Energy Data Book*, 30th Edition, US Department of Energy, ORNL-6986, June 2011.

3. George A. Olah, Alain Goeppert, and G.K. Surya Prakash, *Beyond Oil and Gas: The Methanol Economy*, Second updated and enlarged edition (Wiley-VCH Verlag GmbH & Co. KGaA, Weinhiem, 2009). pg. 110

4. William Pentland, "Energy-Positive Buidings Inch Closer to Mainstream Markets," *Forbes Online*, May 27, 2011; http://www.forbes.com/sites/williampentland/2011/05/27/energy-positive-buildings-inch-closer-to-mainstream-markets/

5. "Volkswagen powers up solar park at Tennessee Passat plant," *Reuters News Service*, 24 January 2013, http://www.reuters.com/article/2013/01/24/us-volkswagen-solar-idUSBRE90N12U20130124.

6. Vaclav Smil, *Energy in Nature and Society: General Energetics of Complex Systems* (The MIT Press, 2008) pg. 284

7. *Worldwide Trends in Energy Use and Efficiency: Key Insights from IEA Indicator Analysis*, International Energy Agency, Paris, 2008.

8. Jos G.J. Olivier, et al., *Trends in global CO2 emissions: 2013 Report* (PBL Netherlands Environmental Assessment Agency, The Hague, 2013),

PBL publication number: 1148, JRC Technical Note number: JRC83593, EUR number: EUR 26098 EN.

9. Jeff Tsao, Nate Lewis, George Crabtree, *Solar FAQs - Sandia National Laboratories*, 2006, US Department of Energy, http://www.sandia.gov/~jytsao/Solar%20FAQs.pdf

10. Wikipedia contributors, "Gansu Wind Farm," Wikipedia, The Free Encyclopedia, http://en.wikipedia.org/wiki/Gansu_Wind_Farm (accessed 19 January 2015).

11. European Wind Energy Association, http://www.ewea.org/policy-issues/offshore/

12. You might wonder how the blade can move faster than the wind if it is being pushed by the wind. Modern wind turbines use aerodynamic lift, the same force used to lift an airplane wing. An engine powers an airplane wing forward during takeoff, and if it goes fast enough the wing generates upward lift due to its shape and angle. The lift is at right angles to the direction of motion, and as long as it keeps moving forward it can stay up in the air. In a wind turbine, the blade is angled relative to the wind to generate lift on the blade, pushing the blade perpendicular to the wind. Since the blade is turning perpendicular to the wind and not running downwind with it, the blade tips can go faster than the wind.

13. Elizabeth Rosenthal, "Tweety Was Right: Cats Are a Bird's No. 1 Enemy," *New York Times*, 20 March 2011, http://www.nytimes.com/2011/03/21/science/21birds.html.

14. Scott R. Loss, Tom Will, and Peter P. Marra, "The impact of free-ranging domestic cats on wildlife of the United States," *Nature Communications* 4:1396 doi: 10.1038/ncomms2380 (2012).

15. Damian Carrington, "High-speed Euro train gets green boost from two miles of solar panels," *The Guardian*, 6 June 2011.

16. John Newman, et al., "Review: An Economic Perspective of Liquid Solar Fuels," *Journal of the Electrochemical Society* vol 159 (10) A1722-A1729 (2012).

17. Timothy A. Wise, "The Cost to Mexico of U.S. Corn Ethanol Expansion" Global Development and Environment Institute Working Paper No. 12-01, Tufts University, May 2012

18. "U.S. Drought 2012: Farm and Food Impacts," United States Department of Agriculture Economic Research Service, http://www.ers.usda.gov/topics/in-the-news/us-drought-2012-farm-and-

food-impacts.aspx, February 2013. The 2012 drought reduced US corn crops by 13% and drove up corn prices by over 20%.

19. Dan Seif, "Advanced, Non-Food Biofuels Come of Age," Rocky Mountain Institute, January 2013, http://blog.rmi.org/ blog_2013_01_22_Advanced_NonFood_Biofuels.

20. Ann R. Thryft, "Update on 100-Percent Non-Food Jet Biofuel," *Design-News*, http://www.designnews.com/author.asp?section_id=1392&doc_id=257969.

21. Vasilis Fthenakis, et al., "The technical, geographical, and economic feasibility for solar energy to supply the energy needs of the US," *Energy Policy* 37 (2009) 387–399.

22. Janet L. Sawin and William R. Moomaw, *Renewable Revolution: Low-Carbon Energy by 2030*, Worldwatch Institute, 2009.

23. Stephen J. DeCanio and Anders Fremstad, "Economic feasibility of the path to zero net carbon emissions," *Energy Policy* 39 (2011) pp 1144–1153.

24. Mark Z. Jacobson and Mark A. Delucchi, "Providing all global energy with wind, water, and solar power, Part I: Technologies, energy resources, quantities and areas of infrastructure, and materials," *Energy Policy* (2010) doi:10.1016/j.enpol.2010.11.040.

25. "State Wind Energy Statistics: Iowa," American Wind Energy Association, 10 April 2014, http://www.awea.org/Resources/state.aspx?ItemNumber=5224

Chapter 13: Energy Storage

1. Steven Chu. [2010]. In Facebook page. Retrieved November 8, 2014, from https://www.facebook.com/notes/steven-chu/why-we-need-more-nuclear-power/336162546856

2. Cory Budischak, et al., "Cost-minimized combinations of wind power, solar power and electrochemical storage, powering the grid up to 99.9% of the time," *Journal of Power Sources* 225 (2013) 60e74.

3. California Public Utilities Commission, "CPUC Sets Energy Storage Goals for Utilities," Press Release 17 October, 2013, Docket # R. 10-12-007, http://docs.cpuc.ca.gov/PublishedDocs/Published/G000/ M079/K171/79171502.pdf.

4. Hal Hodson, "Greening the Grid," *New Scientist*, 2 Feb 2013.

5. Bernard Lee, David Gushee, *Massive Electricity Storage*, American Institute of Chemical Engineers, 2008.

6. Wikipedia contributors, "Renewable energy in Germany" Wikipedia, The Free Encyclopedia, http://en.wikipedia.org/wiki/Renewable_energy_in_Germany (accessed 24 March 2015).

7. *EPRI-DOE Handbook of Energy Storage for Transmission &Distribution Applications*, EPRI, Palo Alto, CA, and the U.S. Department of Energy, Washington, D.C.: 2003. 1001834.

8. Jim Giles, "Staying Power," *New Scientist*, Issue 2898, 5 January 2013.

9. Duncan Graham-Rowe, "Hybrid cars are poised to give flywheels a spin," *NewScientist*, 11 June 2011.

10. A lead-acid battery can store about 120 joules per gram of battery weight, while a lithium battery can store about 900 joules per gram.

11. Yet-Ming Chiang, et al., "Semi-Solid Lithium Rechargable Battery," *Advanced Energy Materials* , 2011, XX, 1–6.

12. Wikipedia contributors, "Fuel cell," Wikipedia, The Free Encyclopedia, http://en.wikipedia.org/wiki/Fuel_cell (accessed 24 July 2014).

13. *FTA Fuel Cell Bus Program: Research Accomplishments through 2011*, FTA Report No. 0014, US Department of Transportation, March 2012.

14. J. M. Duthie and H. W. Whittington, *Securing Renewable Energy Supplies Through Carbon Dioxide Storage in Methanol*, Power Engineering Society Summer Meeting, 2002 IEEE (Volume:1), ISBN 0-7803-7518-1, http://ieeexplore.ieee.org/xpl/articleDetails.jsp?reload=true&arnumber=1043200

15. George A. Olah, Alain Goeppert, and G.K. Surya Prakash,*Beyond Oil and Gas: The Methanol Economy*, Second updated and enlarged edition (Wiley-VCH Verlag GmbH & Co. KGaA, Weinhiem, 2009).

16. "Renewable Methanol," *Milestones 2011*, Methanol Institute, Alexandria VA. 2011.

Chapter 14: Energy Efficiency

1. Vaclav Smil, *Energy in Nature and Society: General Energetics of Complex Systems* (The MIT Press, 2008) pg. 265

2. Amory B. Lovins, *Energy End-Use Efficiency*, Rocky Mountain Institute, Snowmass CO, 19 September 2005.

3. *Energy = Future: Think Efficiency*, American Physical Society, College Park, MD, September 2008.

4. *Energy = Future: Think Efficiency*, American Physical Society, College Park, MD, September 2008.

5. Amory B. Lovins, *Energy End-Use Efficiency*, Rocky Mountain Institute, Snowmass CO, 19 September 2005.

6. *Energy = Future: Think Efficiency*, American Physical Society, College Park, MD, September 2008.

7. Evan Mills, *Building Commissioning: A Golden Opportunity for Reducing Energy Costs and Greenhouse Gas Emissions*, Lawrence Berkeley National Laboratory Berkeley, California, 2009, http://cx.lbl.gov/2009-assessment.html

8. Amory B. Lovins and L. Hunter Lovins, *Climate: Making Sense and Making Money*, Rocky Mountain Institute, 1997

9. Amory B. Lovins, *Energy End-Use Efficiency*, Rocky Mountain Institute, Snowmass CO, 19 September 2005.

10. *Energy = Future: Think Efficiency*, American Physical Society, College Park, MD, September 2008.

11. *International Energy Statistics: Total Primary Energy Consumption per capita*, US Energy Information Administration, http://www.eia.gov/cfapps/ipdbproject/iedindex3.cfm?tid=44&pid=45&aid=2&cid=regions&syid=2011&eyid=2011&unit=MBTUPP, accessed 8 August 2014.

12. Human Development Index (HDI), United Nations Development Program, http://hdr.undp.org/en/content/human-development-index-hdi

13. Wikipedia contributors, "Genuine Progress Indicator," Wikipedia, The Free Encyclopedia,http://en.wikipedia.org/wiki/Genuine_progress_indicator (accessed 22 August 2014).

14. Vaclav Smil, *Energy in Nature and Society: General Energetics of Complex Systems* (The MIT Press, 2008) p 384

Chapter 15: Solar PV Will Be the Biggest Solar

1. Tidal energy and geothermal energy are considered renewable energy but are not forms of solar energy. They may contribute to the future energy mix, but they each have highly specific siting requirements that will limit their spread.

2. Joel Jean, R. Brown, Robert L. Jaffe, Tonio Buonassisi and Vladimir Bulović, *Pathways for solar photovoltaics*, Energy and Environmental Science, 2015, Advance Article DOI: 10.1039/C4EE04073B.

3. Andrew S. Goudie (Editor) and David J. Cuff (Editor), *Encyclopedia of Global Change: Environmental Change and Human Society*, Oxford University Press, 2001, pages 229-230.

4. *Energy Statistics Yearbook 1996*, United Nations Statistics Division, Department of Ecoomic and Social Affairs, 1996.

5. Helmut Haberl, et al., "Quantifying and mapping the human appropriation of net primary production in earth's terrestrial ecosystems," *Proceedings of the National Academy of Sciences* vol. 104 no. 31 (2007), pp 12942–12947.

6. Ottmar Edenhofer, et al., *Special Report on Renewable Energy Sources and Climate Change Mitigation*, IPCC Special Report, (Cambridge University Press, 2011). pg. 3-52

7. Mark Z. Jacobson and Mark A. Delucchi, "Providing all global energy with wind, water, and solar power, Part I: Technologies, energy resources, quantities and areas of infrastructure, and materials," *Energy Policy* (2010) doi:10.1016/j.enpol.2010.11.040.

8. Ottmar Edenhofer, et al., *Special Report on Renewable Energy Sources and Climate Change Mitigation*, IPCC Special Report, (Cambridge University Press, 2011). pg. 3-51.

9. SEMI, http://www.semi.org

10. SEMI, http://www.semi.org/en/Standards/CTR_035945

11. A training program for firefighters to handle photovoltaic systems is available from the California Office of the State Fire Marshal, http://osfm.fire.ca.gov/training/photovoltaics.php.

12. Wikipedia contributors, "Coal mining," Wikipedia, The Free Encyclopedia, http://en.wikipedia.org/wiki/Coal_mining (accessed 22 August 2014).

13. *Oil in the Sea III: Inputs, Fates, and Effects*, US National Research Council, National Academies Press, 2003.

14. Wikipedia contributors, "Petroleum," Wikipedia, The Free Encyclopedia, http://en.wikipedia.org/wiki/Petroleum (accessed 28 July 2014).

15. Jordan Macknick, et al., *A Review of Operational Water Consumption and Withdrawal Factors for Electricity Generating Technologies*, US National Renewable Energy Laboratory, Technical Report NREL/TP-6A20-50900, March 2011.

16. James Kanter, "Climate change puts nuclear energy into hot water," *New York Times*, 20 May 2007, http://www.nytimes.com/2007/05/20/health/20iht-nuke.1.5788480.html

17. Jordan Macknick, et al., *A Review of Operational Water Consumption and Withdrawal Factors for Electricity Generating Technologies*, US National Renewable Energy Laboratory, Technical Report NREL/

TP-6A20-50900, March 2011. The median water usage is 687 gallons per megawatt-hour for a tower-cooled coal power plant, and 672 for a tower-cooled nuclear power plant.

18. Michael Liebreich, "Water May Top Up the Case for Renewables," *Bloomberg New Energy Finance*, 27 September 2012, http://bit.ly/S1Jjk1.

19. Benjamin Haas, "China Misses Output Targets as It Envies U.S. Shale Gas Success," *Bloomberg News*, 20 February 2014, http://www.bloomberg.com/news/2014-02-20/china-misses-output-targets-as-it-envies-u-s-shale-gas-success.html.

20. Jordan Macknick, et al., *A Review of Operational Water Consumption and Withdrawal Factors for Electricity Generating Technologies*, US National Renewable Energy Laboratory, Technical Report NREL/TP-6A20-50900, March 2011.

21. Plymouth Area Renewable Energy Initiative (PAREI), http://www.plymouthenergy.org/

22. Sustainable Development For All, http://sustainabledevelopmentforall.org/

23. Reese Rogers, "Smart Grid and Energy Storage Installations Rising," *Vital Signs Online*, Worldwatch Institute, February 26, 2013.

24. Hawai'i Clean Energy Initiative. [2015]. Retrieved on February 25, 2015 from http://www.hawaiicleanenergyinitiative.org/

25. Roger H. Bezdek, *Estimating the Jobs Impacts of Tackling Climate Change*, American Solar Energy Society, October 2009.

26. *U.S. International Trade in Goods and Services*, U.S. Census Bureau, U.S. Bureau of Economic Analysis, August 2011.

27. *How much land will PV need to supply our electricity?*, PV FAQs, US Department of Energy, http://www.nrel.gov/docs/fy04osti/35097.pdf.

28. R. Carbone, et al., *Photovoltaic systems for powering greenhouses*, 2011 International Conference on Clean Electrical Power (ICCEP), 14-16 June 2011.

29. Amory Lovins, "Mighty Mice: The most powerful force resisting new nuclear may be a legion of small, fast and simple microgeneration and efficiency projects," *Nuclear Engineering International*, December 2005, http://www.neimagazine.com.

30. Silicon Valley Toxics Coalition, http://svtc.org

31. PV CYCLE Association, http://pvcycle.org

32. Colleen McCann Kettles, "A Comprehensive Review of Solar Access Law in the United States: Suggested Standards for a Model Statute and Ordinance," Solar America Board for Codes and Standards, 2008, http://www.solarabcs.org/solaraccess.

Chapter 16: The Compelling Economics of Solar PV

1. PVWatts calculator available online at http://www.nrel.gov/rredc/pvwatts/

2. The Pacific Gas & Electric website shows that the average monthly electricity bill for low users in March 2015 was $89.30 for 500 kwh. That works out to 17.9¢ per kwh. http://www.pge.com/en/myhome/saveenergymoney/plans/rateupdates/about/grc.page

3. Ben Hoen, et al., *An Analysis of the Effects of Residential Photovoltaic Energy Systems on Home Sales Prices in California*, LBNL-4476E, Lawrence Berkeley National Laboratory, April 2011.

4. Samuel R. Dastrup, et al., "Understanding the Solar Home price premium: Electricity generation and 'Green' social status," *European Economic Review* vol. 56(5), 2012, pages 961–973.

5. Nichoas Stern, *Stern Review on The Economics of Climate Change* (HM Treasury, London, 2006).

6. Wikipedia contributors, "Stern Review," Wikipedia, The Free Encyclopedia, http://en.wikipedia.org/wiki/Stern_Review (accessed 22 July 2014).

7. *Hidden Costs of Energy: Unpriced Consequences of Energy Production and Use*, Committee on Health, Environmental, and Other External Costs and Benefits of Energy Production and Consumption, National Research Council, National Academies Press, 2010.

8. Paul R. Epstein, et al., "Full cost accounting for the life cycle of coal," *Annals of the New York Academy of Sciences* 1219 (2011) 73–98.

9. *Imported Oil and U.S. National Security*, RAND Corporation, http://www.rand.org, 2009.

10. Richard Perez, et al., "Solar Power Generation in the US: Too expensive, or a bargain?" *Energy Policy* vol 39, Issue 11, November 2011, pages 7290–7297

11. Some efforts are being made to monetize solar value. The state of Minnesota developed a methodology for valuing solar energy to justify their Solar Tariff. See http://mn.gov/commerce/energy/topics/resources/energy-legislation-initiatives/value-of-solar-tariff-methodology%20.jsp

Chapter 17: The Price of Solar PV is Dropping

1. "Solar surge drives record clean energy investment in 2011," Bloomberg New Energy Finance, Press Release 12 January 2012.

2. *Solar generation 6: Solar photovoltaic electricity empowering the world*, European Photovoltaic Industry Association and Greenpeace, 2011.

3. "Daily Chart: Pricing Sunshine," Economist,com, 28 December 2012, http://www.economist.com/blogs/graphicdetail/2012/12/daily-chart-19, accessed 28 July 2014.

4. Ottmar Edenhofer, et al., *Special Report on Renewable Energy Sources and Climate Change Mitigation*, IPCC Special Report, (Cambridge University Press, 2011).

5. Ottmar Edenhofer, et al., *Special Report on Renewable Energy Sources and Climate Change Mitigation*, IPCC Special Report, (Cambridge University Press, 2011).

6. Ottmar Edenhofer, et al., *Special Report on Renewable Energy Sources and Climate Change Mitigation*, IPCC Special Report, (Cambridge University Press, 2011).

7. Ottmar Edenhofer, et al., *Special Report on Renewable Energy Sources and Climate Change Mitigation*, IPCC Special Report, (Cambridge University Press, 2011).

8. Mingzhen Liu, et al., "Efficient planar heterojunction perovskite solar cells by vapour deposition," *Nature* vol 501, pp 395–398, 19 September 2013.

9. Wikipedia contributors, "Multijunction photovoltaic cell," Wikipedia, The Free Encyclopedia, http://en.wikipedia.org/wiki/Multijunction_photovoltaic_cell (accessed 23 July 2014).

10. 1366 Technologies, Inc., http://1366tech.com/technology/direct-wafer/

11. Martin LaMonica, "50 Smartest Companies: 1366 Technologies," *MIT Technology Review*, Massachusetts Institute of Technology, March/April 2014.

12. K.R. Catchpole and A. Polman, "Plasmonic Solar Cells," *Optics EXPRESS*, 22 December 2008, vol 16, no. 26, pg. 21793.

13. Alexander Ip, et al., "Hybrid passivated colloidal quantum dot solids," *Nature Nanotechnology*, vol 7, no. 9, pp. 577–582 (2012).

14. Dale K. Kotter, et al., "Solar Nantenna Electromagnetic Collectors," *Proceedings of Energy Sustainability 2008*, August 10-14, 2008, Jacksonville FL, ES2008–54016.

15. Dexter Johnson, "Nano-antenna Arrays May Yield Ultra-Efficient Solar Devices," *IEEE Spectrum*, 6 February 2013, http://spectrum.ieee.org/nanoclast/semiconductors/nanotechnology/nanoantenna-arrays-may-yield-ultraefficient-solar-devices

Chapter 18: How Much Solar is Needed?

1. *The Copenhagen Diagnosis 2009: Updating the World on the Latest Climate Science*, University of New South Wales Climate Change Research Centre, Sydney, Australia, 2009.

2. Stephen J. DeCanio and Anders Fremstad, "Economic feasibility of the path to zero net carbon emissions," *Energy Policy* 39 (2011) pp 1144–1153.

3. Peter D. Schwartzman and David W. Schwartzman, *A Solar Transition is Possible*, The Institute for Policy Research & Development (IPRD), London, March 2011.

4. Mark Z. Jacobson and Mark A. Delucchi, "Providing all global energy with wind, water, and solar power, Part I: Technologies, energy resources, quantities and areas of infrastructure, and materials," *Energy Policy* (2010) doi:10.1016/j.enpol.2010.11.040.

5. *Energy [R]evolution: A Sustainable World Energy Outlook*, Greenpeace International and European Renewable Energy Council, 3rd edition, June 2010.

6. *Climate 2030: A National Blueprint for a Clean Energy Economy*, Union of Concerned Scientists, Cambridge, MA, 2009.

7. *World Population Prospects, the 2010 Revision*, United Nations, Department of Economic and Social Affairs.

8. *International Energy Statistics: Total Primary Energy Consumption per capita*, US Energy Information Administration, http://www.eia.gov/cfapps/ipdbproject/iedindex3.cfm?tid=44&pid=45&aid=2&cid=regions&syid=2011&eyid=2011&unit=MBTUPP, accessed 8 August 2014.

9. Janet L. Sawin and William R. Moomaw, *Renewable Revolution: Low-Carbon Energy by 2030*, Worldwatch Institute, 2009.

10. Peter D. Schwartzman and David W. Schwartzman, *A Solar Transition is Possible*, The Institute for Policy Research & Development (IPRD), London, March 2011.

11. Nathan S. Lewis, "Powering the Planet," *Engineering & Science* No. 2, 2007. pg. 18

12. Sean Ong, et al., *Land-Use Requirements for Solar Power Plants in the United States*, US National Renewable Energy Laboratory, Technical Report NREL/TP-6A20-56290, June 2013.

13. Vaclav Smil, *Energy in Nature and Society: General Energetics of Complex Systems* (The MIT Press, 2008) pg. 382

14. Vaclav Smil, *Energy in Nature and Society: General Energetics of Complex Systems* (The MIT Press, 2008) pg. 382

15. *Electric Power Monthly: with Data for April 2014*, US Energy Information Administration, June 2014.

16. *How much land will PV need to supply our electricity?*, PV FAQs, US Department of Energy, http://www.nrel.gov/docs/fy04osti/35097.pdf.

17. Navigant Consulting, Inc., *California Rooftop Photovoltaic (PV) Resource Assessment and Growth Potential by County*, California Energy Commission, PIER Program. CEC-500-2007-048, 2007.

18. *MRS Bulletin*, vol 33, issue 04, April 2008, pp 456–458, Materials Research Society. Published online by Cambridge University Press, January 2011, DOI: http://dx.doi.org/10.1557/mrs2008.90

Chapter 19: Exponential Growth

1. European Photovoltaic Industry Assocation, *Global Market Outlook for Photovoltaics 2014-2018*, EPIA, Brussels, Belgium, 2014.

2. "PV Capacity Expansion to Slow to 10% in 2012; 2011 Expansion Tops 50%," IMS Research press release, November 3, 2011, http://www.pvmarketresearch.com

3. European Photovoltaic Industry Assocation, *Global Market Outlook for Photovoltaics 2014-2018*, EPIA, Brussels, Belgium, 2014.

4. For a mathematical explanation of this rule, see the article titled "Rule of 72" in Wikipedia.

5. *Cell Phone Subscribers in the U.S., 1985–2010*, InfoPlease, http://www.infoplease.com/ipa/A0933563.html

6. "Global Market Outlook for Photovoltaics until 2016," European Photovoltaic Industry Association, 2012.

7. Richard Perez, et al., "Solar Power Generation in the US: Too expensive, or a bargain?" *Energy Policy* vol 39, Issue 11, November 2011, pages 7290–7297

8. Richard Perez, et al., "Solar Power Generation in the US: Too expensive, or a bargain?" *Energy Policy* vol 39, Issue 11, November 2011, pages 7290–7297

9. Jamie Speirs, et al., *Materials Availability: Potential constraints to the future low-carbon economy*, UK Energy Research Centre, April 2011: REF UKERC/WP/TPA/2011/002.

Chapter 20: Electrification

1. Institute for Sustainable Development and International Relations, *Pathways to Deep Decarbonization: Interim 2014 Report*, Published by Sustainable Development Solutions Network and Institute for Sustainable Development and International Relations, July 2014, http://deepdecarbonization.org.

2. *Annual Energy Review 2010*, Energy Information Administration, US Department of Energy, DOE/EIA-0384(2010), October 2011.

3. Timothy Searchinger, "A Quick Fix to the Food Crisis," *Scientific American*, July 2011.

4. George A. Olah, Alain Goeppert, and G.K. Surya Prakash,*Beyond Oil and Gas: The Methanol Economy*, Second updated and enlarged edition (Wiley-VCH Verlag GmbH & Co. KGaA, Weinhiem, 2009). pg. 118

5. Rob van Haaren, *Assessment of Electric Cars Range Requirements and Usage Patterns based on Driving Behavior recorded in the National Household Travel Survey of 2009*, Earth and Environmental Engineering Department, Columbia University, December 2011, http://www.solarjourneyusa.com/EVdistanceAnalysisFullText.php

6. Steve Raabe, "Colorado-based UQM to supply motors for 100 all-electric UPS trucks," *Denver Post*, 5 January 2012.

7. Jerry Hirsch, "Postal Service asks five firms to help it deliver in a green way," *Los Angeles Times*, 16 Februrary 2010.

8. Wikipedia contributors, "Electric aircraft," Wikipedia, The Free Encyclopedia, http://en.wikipedia.org/wiki/Electric_aircraft (accessed 19 August 2014).

9. James H. Williams, et al., "The Technology Path to Deep Greenhouse Gas Emissions Cuts by 2050: The Pivotal Role of Electricity," *Science*, 6 January 2012: vol. 335 no. 6064 pp. 53–59.

10. Gary B. Nash (Editor), *Encyclopedia of American History, Volume 7: The Emergence of Modern America, 1900 to 1928*, Facts on File, 2002.

11. *Electricity's share of U.S. delivered energy has risen significantly since 1950*, US Energy Information Administration, Annual Energy Review, March 2, 2012, http://www.eia.gov/todayinenergy/detail.cfm?id=5230, accessed 10 August 2014. These percentages leave out transportation, which is mostly petroleum-based.

Chapter 21: Transition Gradually and Gracefully

1. Amory Lovins, et al., "Nuclear Power: Climate Fix or Folly?," *RMI Solutions*, April 2008, Rocky Mountain Institute.

Chapter 24: What Holds Us Back?

1. Rankings by Fortune Magazine based on company revenue. See http://fortune.com/global500/2013/.

2. Intergovernmental Panel on Climate Change, *Climate Change 2013: The Physical Science Basis. Contribution of Working Group I to the Fifth Assessment Report of the Intergovernmental Panel on Climate Change* [Stocker, T.F., D. Qin, G.-K. Plattner, M. Tignor, S.K. Allen, J. Boschung, A. Nauels, Y. Xia, V. Bex and P.M. Midgley (eds.)] (Cambridge University Press, Cambridge, United Kingdom and New York, NY, USA).

3. *Smoke, Mirrors & Hot Air*, Union of Concerned Scientists, January 2007

4. *Koch Industries Still Fueling Climate Denial*, 2011 Update, Greenpeace USA, Washington D.C., April 2011.

5. *Oil and Gas Lobbying*, Center for Responsive Politics website, based on records in the Senate Office of Public Records, 2011.l

6. Roger Pielke, Jr., *The Climate Fix: What Scientists and Politicians Won't Tell You About Global Warming* (Basic Books, 2010)

7. *CO2 Emissions from Fuel Combustion:Highlights*, 2013 Edition, International Energy Agency, Paris, France.

8. *Annual Energy Review 2009*, Energy Information Administration, US Department of Energy, DOE/EIA-0384(2009), August 2010.

9. *Estimating U.S. Government Subsidies to Energy Sources: 2002-2008*, Environmental Law Institute, Washington D.C., September 2009.

10. In 2010, the Iranian Parliament voted to gradually scale back the subsidies over a five-year period.

11. *World Energy Outlook 2013: Executive Summary*, International Energy Agency, Paris, France, 2013.

12. Congressional Record—House, Vol. 153, Pt. 2, pg. 1598, January 18, 2007.

13. David McCollum et al., "Energy Investments Under Climate Policy: A Comparison of Global Models," *Climate Change Economics*, vol 04, issue 04, November 2013, DOI 10.1142/S2010007813400101.

14. *Renewables 2011 Global Status Report*, Renewable Energy Policy Network for the 21st Century, Paris, 2011. pg. 14.

15. Wikipedia contributors, "Carbon emission trading," Wikipedia, The Free Encyclopedia, http://en.wikipedia.org/wiki/Carbon_emission_trading (access 22 July 2014).

16. *America's Climate Choices*, Committee on America's Climate Choices, National Research Council (National Academies Press, 2011).

17. *ExxonMobil's views and principles on policies to manage long-term risks from climate change*, http://corporate.exxonmobil.com/en/current-issues/climate-policy/climate-policy-principles/overview, accessed 12 March 2015.

18. Citizens' Climate Lobby, http://citizensclimatelobby.org/

19. Patrick Foulis, "British Columbia's carbon tax: The evidence mounts," The Economist website, , 31 July 2014, http://www.economist.com/blogs/americasview/2014/07/british-columbias-carbon-tax (accessed 12 September 2014).

20. *Solar generation 6: Solar photovoltaic electricity empowering the world*, European Photovoltaic Industry Association and Greenpeace, 2011.

21. *Global Market Outlook for Photovoltaics until 2015*, European Photovoltaic Industry Association, 2011.

22. *Renewable Electricity Generation in Germany*, Erneuerbare-Energien-und-Klimaschutz.de, volker-quaschning.de, 2015, http://www.volker-quaschning.de/datserv/ren-Strom-D/index_e.php.

23. "Support program for Solar Power Storage Systems to begin May 1st," Bundesverband Solarwirtschaft (German Solar Industry Association), 18 April 2013, http://www.solarwirtschaft.de/en/media/single-view/news/support-program-for-solar-power-storage-systems-to-begin-may-1st.html

24. Tariff Watch, PV-Tech, http://www.pv-tech.org/tariff_watch/list, accessed 9 March 2015

25. *Energy [R]evolution: A Sustainable World Energy Outlook*, Greenpeace International and European Renewable Energy Council, 3rd edition, June 2010.

26. Jeremy Carl (Editor), James E. Goodby (Editor), *Conversations about Energy: How the Experts See America's Energy Choices*, Hoover Institution Press; 1st edition (November 13, 2010), pg. 24

Chapter 25: The Way Forward: Individual Action

1. *The Solar America Cities Awards*, a Solar America Initiative Fact Sheet, US Department of Energy, DOE/GO-102008-2586, March 2008.

2. *Organic Food Poll*, Consumer Reports National Research Center, Consumers Union of U.S., Inc., February 8, 2010 NRC #2010.13.

3. Amory B. Lovins, *Energy End-Use Efficiency*, Rocky Mountain Institute, Snowmass CO, 19 September 2005.

4. *America's Climate Choices*, Committee on America's Climate Choices, National Research Council (National Academies Press, 2011).

Chapter 26: Time to Act

1. The PVWatts solar calculator provided by the US National Renewable Energy Laboratory is available at http://www.nrel.gov/rredc/pvwatts/. Entering a 4 kilowatt PV array tilted up to 20 degrees and facing due south in Phoenix, Arizona, the calculator predicts the system will produce 6, 919 kilowatt-hours per year. Entering the same design for Seattle, Washington, yields 4,362 kilowatt-hours per year, a difference of 58.6%.

2. http://solarpanelhost.org/garden/california/pvusa

3. Eugene Buchanan, "Growing Solar in Your Community," *Home Power Magazine*, 143, June & July 2011.

4. Westmill Solar Coop, http://www.westmillsolar.coop.

5. Mosaic, 426 17th Street, 6th floor Oakland, CA 94612, 510-746-8602, http://joinmosaic.com

6. Everybody Solar, 362A 11th Avenue, San Francisco, CA 94118, www.everybodysolar.org

7. RE-volv, 972 Mission St., Suite 500, San Francisco, CA 94103, http://www.re-volv.org

Index